Integrated Design and Engineer
as a Business Improvement Proce:

Integrated Design and Engineering

as a Business Improvement Process

ir. T.M.E. Zaal

Maj Engineering Publishing

Colofon

Title: Integrated Design and Engineering -
                       as a Business Improvement Process

Author: ir. T.M.E. Zaal

Editor: Steve Newton

Publisher: Maj Engineering Publishing

ISBN: 978 90 79182 03 9

Edition: First edition, First impression, February 2009

Cover design: Carlito's Design

Design & Layout: CO2 Premedia

Cartoons: W. Rietkerk, wr.cartoons.nl

©2009 Maj Engineering Publishing

Foreword

Developing innovations for new products, services and processes is always a challenging and dangerous process. Challenging because innovation means facing new challenges, new ways of operating and new possibilities. On the other hand it is very difficult and even dangerous because during the development process decisions and choices have to continuously take place which exclude some of these possibilities. All the decisions and choices being made during this process will have great impact on the cost price of the new product and, therefore, also on the future profitability of a company. So, innovation through developing a new product or service is not only challenging but also a very risky business. Everybody knows examples of products that were not good enough to achieve profitability in their market. Sometimes companies even face bankruptcy through poor products. Good design equates to lots of profit, poor design equates to lots of troubles.

This book Integrated Design and Engineering originates from several projects between universities and industries in The Netherlands that focused on how to innovate the design and engineering processes. In addition, the research work undertaken by the chair of Integrated Design at the Hogeschool Utrecht University of Applied Science has provided significant input for the content of this book, as have the workshops, lectures and courses, about this subject. Thanks are due to the members of staff of this chair: Frans Speelman, Geert-Jan Temmink, Marc Pallada, Kees Dingemans, Pim van der Waal, Bert van Huygevoort, Henk Burggraaff and Robert Huls for their continuous and critical inputs on Integrated Design. I would also like to thank Alexander Udink ten Cate for his inspiration and support to start up this chair and Robert Blom for his institutional help funding it.

From the industrial world I would like to thank Theo Lohman, Herman Eekels and Jan Hak for their enthusiastic input and activities in promoting the ideas of Integrated Design in the Netherlands. On numerous occasions we have jointly given promotional speeches in industrial settings about the requirement to introduce the concepts of improvement to the design processes in industrial organizations.

Finally, I thank my wife Ine for the endless patience and encouragement she has given me, whilst somewhere in our house the creation process for this book was taking place.

This book about Integrated Design and Engineering will focus on the development process in a company as a pure business process, with the main goal being to develop profitable products and services. The central issue in this development process is how to realize 'creating added value' both for clients and for organizations.

What lessons will you learn after reading and, in particular, applying this book:

- What is involved in setting up a design and engineering process that is client oriented and value driven for your organization.
- How to organize an improvement of existing products and services with all the stakeholders.
- How to implement the role of information technology over the whole life cycle of a product, including the reuse of proven knowledge.
- Exciting applications from the fields of designing products, building services and asset management.

Tim Zaal,
Emeritus Associate Professor
Utrecht University of Applied Science, The Netherlands
Hoorn, February 2009

Brief content description

Chapter 1: Why an Integrated Design and Engineering Process?
An introduction to the content of the Integrated Design and Engineering (IDE) philosophy and the drivers to apply this new way of thinking to the product development process. By introducing this way of working to organize the development process as a business process, an organization has to be aware that it will have consequences for cooperation between departments in the organization. It also provides an overview of the benefits of applying this way of working.

Chapter 2: IDE Philosophy Demands for Team Working by Concurrent Engineering
The development team is always a crucial part in the successful development of a new product. The composition of the team is of great importance so we should take a lot of care in putting together the right team. When this team applies the IDE philosophy then there is a good chance of completing the process with the right product. The starting points of the concurrent engineering principles for the team work are introduced as the way to organize and plan the development work. The Lean principles are applied too for the lay-out of the process. An overview of IDE competences is also given.

Chapter 3: IDE Structure Models for Product Data Management and Reuse of Knowledge
In the Integrated Design and Engineering philosophy the integration of information is a main issue. Storing of all types of data in a structured way in an engineering database is essential. Some structural models are presented that are able to facilitate the set up of such a database. Advantages of such a database include the ability to enter failure free data and the reuse of proven knowledge.

Chapter 4: How to Map Client Demands and Wishes
An introduction to using processes to map ideas for innovations and new products that reflect client demands and wishes. Methods like Systems Engineering, Functional Specifications and Quality Function Deployment can be applied by the integrated design team to develop a product definition document with all these demands and wishes weighted and selected.

Chapter 5: Methodology for Product Design over the Life Cycle

On basis of the product definition document, a new product can be developed by the integrated design team using the methodologies of structural designing. A number of methodologies are available. The result has to be a product proposal that promises the best possible price-quality, best added value, highest value of sustainability and lowest life cycle costs.

Chapter 6: Production, Operate, Maintain and Continuous Improvement

Frequently more than 90% of the work of a development department is improvement work on existing products or services. A continuous review of existing products and manufacturing processes to identify possible improvements is very useful and, for the profitability of a company, can often be more important than totally new concepts that have a high degree of risk. Designing for long life service provides a company with opportunities in relation to maintenance services and continuous improvements through modification kits over the whole life cycle of a product or service. This long life service concept generates a constant income during this time. Well known improvement methodologies like Lean, Six Sigma and Value Analysis are applied in the IDE philosophy.

Chapter 7: Maintenance and Operation

The design of a product and, through this, the requirement for repair (or maintenance) activities has to be optimized in order to achieve optimal results in the future operation of a company. A methodology is presented to check existing and new products or installations in relation to their maintenance behavior and also to show possibilities for improvements in this field.

Table of Contents

Foreword V
Brief content description VII
Table of Contents IX
List of Abbreviations. XII

1   Why an Integrated Design and Engineering Process?   1

1.1   Introduction. 1
1.2   Business and Profit. 2
1.3   IDE as a Business Process 8
1.4   What is an Integrated Design and Engineering Process? 8
1.5   Role of ICT for IDE in Organizations 11
1.6   Integrated Design and Engineering in a Company 12
Summary. 18
Exercises 18
Literature. 19

**2   IDE Philosophy Demands for Team Working by Concurrent
      Engineering   21**

2.1   Introduction. 21
2.2   What is the Right Team?. 21
2.3   Building the Product Development Team .28
2.4   Concurrent Engineering. 30
2.5   Lean Development Process for Product Development. 31
2.6   Team Work in the Building Environment. .34
2.7   Team Work and Education 34
2.8   Management Paradox (1) 36
Summary. 38
Exercises 38
Literature. 38

**3   IDE Structure Models for Product Data Management
      and Reuse of Knowledge   39**

3.1   Introduction. 39
3.2   Information Systems for Organizations 40
3.3   Structure Models for IDE Applications 43
3.4   Code systems. 50

3.5   Hamburger model as Product Configuration Model (PCM) 55
3.6   Management Paradox (2) 56
Summary. 57
Exercises 57
Literature. 58

4   How to Map the Client's Demands and Wishes   59

4.1   Introduction. 59
4.2   Developing or Creating Business with Clients 61
4.3   Quality Function Deployment (QFD) 68
4.4   Customer Satisfaction and the Kano Model 72
4.5   QFD and the Other Six Chambers 74
4.6   Product Definition Document (PDFDoc) 84
Summary. 86
Exercises 86
Literature. 87

5   Methodology for Product Design over the Life Cycle   89

5.1   Introduction. 89
5.2   Designing over the Whole Life Cycle. 90
5.3   Phases over the Life Cycle 90
5.4   Concurrent Engineering Planning 97
5.5   ICT over the Life Cycle 98
5.6   Costs of Manufacturing, Life Cycle and Investment 99
5.7   Parts Engineering 102
5.8   Design Team and Part Engineering. 107
5.9   Solutions for Modules and Parts 116
5.10   Manufacturing Selection 119
5.11   Design for Manufacturing 120
5.12   Design for Operation and Maintainability 121
5.13   Design for Long Life Service 122
5.14   Design for Sustainability and Reuse 123
5.15   Design for Production Costs and Life Cycle Costs 123
5.16   Design Process as Value Chain Business Model 124
5.17   Cost engineering. 126
5.18   Design Kernel for Engineering 126
5.19   Product Configuration Model for a Product Family. 128
Summary. 132
Exercises 132
Literature. 134

6 **Production, Operation, Maintenance and Continuous Improvement**   **135**

6.1   Introduction. 135
6.2   The After Sales Problems 136
6.3   Improvement Activities 140
6.4   Analysis Methodologies for Eliminating Complaints. 141
6.5   Business Model for Improvements 152
6.6   After Sales Engineering Process 152
Summary. 155
Exercises 155
Literature. 155

7 **Maintenance and Operation**   **159**

7.1   Introduction. 159
7.2   Maintenance as a Process. 160
7.3   Maintenance required through process failure. 163
7.4   Maintenance Repair or Restoration Strategies 166
7.5   Maintenance Strategies for New Products 175
7.6   Functional critical Failure Modes and Effects Analysis (*Fc*-FMEA). . . . 180
7.7   Product Development and Maintenance Plan 189
7.8   Commissioning and Early Equipment Management. 189
7.9   Total Productive Maintenance (TPM) 191
Summary. 197
Exercises 197
Literature. 199

8 **Answers to the problems**   **201**

Chapter 1 201
Chapter 2 204
Chapter 3 205
Chapter 4 206
Chapter 5 209
Chapter 6 212
Chapter 7 215

Index 219

List of Abbreviations

A	Availability
AM	Autonomous Maintenance
CAD	Computer Aided Design
CAM	Computer Aided Manufacturing
CBM	Condition Based Maintenance
CEP	Concurrent Engineering Principles
CM	Criticality Matrix
CMMS	Computerized Maintenance Management System
DCS	Digital Control System
ERP	Enterprise Resource Planning
EQFD	Enhanced Quality Function Deployment
FBM	Failure Based Maintenance
FC	Failure Consequences
FCA	Failure Consequences Analysis
Fc-FMEA	Functional Critical FMEA
FE	Failure Effect(s)
FEM	Failure Eliminating by Modification
FF	Functional Failure
FM	Failure Mode(s)
FME(C)A	Failure Modes and Effects (Criticality) Analysis
FS	Functional Specifications
FTA	Fault Tree Analysis
FU	Function
HoQ	House of Quality
ICT	Information and Communication Technology
IDE	Integrated Design and Engineering
IDEF-o	Integration Definition for Function Modelling
λ	Amount of stops or failures per unit time
λR	Stops by Repair per unit time
λF	Stops by Failures per period of time
LCC	Life Cycle Costs
LCE	Life Cycle Engineering
MTBF	Mean Time Between Failures (Unplanned)
MTBP	Mean Time Between Production Break Downs
MTBR	Mean Time Between Repairs (Planned)
MTPS	Mean Time Production Stops
MTTR	Mean Time To Repair
OEE	Overall Equipment Efficiency
OEM	Original Equipment Manufacturer

PBP	Pay Back Period
PENDoc	Product Engineering Document
PLC	Programmed Logical Controller
PCM	Product Configuration Model
PDM	Product Data Management
PE	Production Efficiency
PLM	Product Life Cycle Management
PM	Planned Maintenance
PMA	Doc Product Manufacturing Document
PPM	Planned Preventive Maintenance
PSPDoc	Product Support Document
Q	Quality rate (or efficiency)
QFD	Quality Function Deployment
R	Reliability
RAM (specs)	Reliability, Availability, Maintainability (specifications)
RCFA	Root Cause Failure Analysis
RCM	Reliability Centered Maintenance
RFB	Repair after Failure Break Down
ROI	Return on Investment
σ	Standard Deviation (Normal Distribution)
SE	Systems Engineering
SO	Solution or Asset that fulfill a FU and the FS
SQC	Substitute Quality Characteristic
TBM	Time Based Maintenance
TCO	Total Costs of Ownership
TDT	Total Dead Time
TOT	Total Operational Time
TPM	Total Productive Maintenance
TPT	Total Productive Time
TRIZ	Theory of Inventive Problem Solving (translated form Russian)
UFM	Unplanned Failure Maintenance
VoC	Voice of the Customer

Chapter 1
Why an Integrated Design and Engineering Process?

1.1   Introduction

Designing new products is always a very risky business for an organization and requires a lot of effort. This effort is in the form of investments, human capital and market orientation, as well as considerable financial resources. On the other hand the competition between companies is very tough, and is becoming increasingly global. The markets ask continuously for new products and for improved performances of existing ones, all at a lower price. This means that organizations have to be very careful how to spend their resources when developing and designing totally new products. Therefore the design activities are very critical for future profits and have to be organized as a business process, with clear goals and time scales.

On the other hand it is known that organizations have a lot of troubles with:

- the integration of design and engineering processes to respond to client wishes;
- the use of ICT systems like CAD-CAM, PDM, ERP, etc, as integrated processes;
- the storing of well proven knowledge of products (parts, techniques and technologies) in databases in a structured, software independent way;
- the reuse of this proven knowledge from these (engineering) databases;
- working together in teams within an organization on basis of concurrent engineering;
- being on time to the market with new products and so making enough profit;
- working together with other organizations on basis of collaborative or simultaneous engineering principles.

We will learn that the Integrated Design and Engineering process (IDE process) can be very helpful in overcoming these problems. It can be applied to both product and process development programs, so when we speak about a product, in practically all cases this process is also applicable for process (improvement) development programs.

Applying this methodology in the real world means that the organization of a company has to fulfill certain conditions, particularly in relation to working in teams and co-operating between departments.

1.2   Business and Profit

For the continuity of a company, a certain minimum amount of profit is necessary. To show the relationship between the product development process and future profit, some business models are presented. We will begin with a simplified business relation model with a relationship between Income, Costs, Profit and Return on Investment (see figure 1.1a). We start with the cash flow, which is income (or turnover) minus expenditures (or costs), or is depreciation plus profit. So profit is the cash flow minus the depreciation. A company has to generate enough cash flow to cover both depreciation and profit. Depreciation funds mean money for investments in ongoing activities, whilst profit means income for new plans and investments, and income for the shareholders. So a company has to generate enough profit so that it can make plans for investments, e.g. developing new products. The **Return on Investment (ROI)** is defined by Profit divided by the Total Value of Investments (in fixed assets).

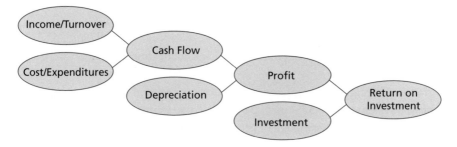

Figure 1.1a Simplified business relation model

Example
If the income is €22 million, the costs are €20 million, than the cash flow is €2 million.

The total value of investments is €10 million, the depreciation is 10% of total value of investment, thus giving €1 million, then the profit is €1 million and the ROI is 1/10 × 100% = 10%.

What is enough profit? Generally speaking it is when the ROI is 3% above the interest rate that investors who are interested in investing in a company in order to finance new plans, including developing new products, would get from state bonds. From the depreciation funds the continuous improvements of existing products can be financed.

The model can also be applied for formulating an innovation or improvement agenda. Firstly we can formulate ideas and plans for decreasing the costs through improvements in production processes. Secondly, we can develop plans for product improvements via extra functionality, better quality, etc. This can give an increase in income, and consequently in cash flow and profit. And thirdly, there can be a development program for totally new products which can also generate extra new cash flow. In the model we can represent these improvements by process innovation of existing processes (lower costs), by product innovation of existing products (higher turn over and lower costs) and by new products (new cash flow), see figure 1.1b.

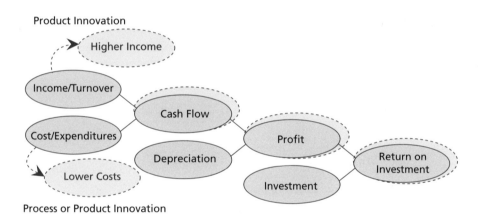

Figure 1.1b Models for innovation or improvement

The innovation or improvement agenda has to be profit driven and can be formulated as follows:

- Production costs or quality improvements: this has to be a continuous process for lowering the production costs. Can be qualified as *Not Risky*;
- Product improvements: these can be divided into two main possibilities:
  - Improvements after client complaints, mostly *Not Risky*;
  - Improvement from new ideas or client questions, mostly *Not Risky*.
- New products: these can also generally divided into two possibilities:
  - New product for an existing market, this can be *Risky*;
  - New product for a new market, this can be *Very Risky*.

In general about 90% of the product development time goes on improvements to existing products, with the rest on risky new products. Why are the new products so risky? Because the chance of going wrong is very real, with the chance of a lot of expensive investment money being lost and not giving future profits.

1.2.1 Relationship between turnover and costs

Another model shows the relationship between the **turnover** of a company and the variable and the **fixed costs**, so we can calculate the point at which this goes from the loss situation to the profit situation (see figure 1.2). When we look at the turnover as a column then we will see at the bottom the variable costs, or in other words all the costs necessary to produce products ready to sell; above these are the fixed costs (including depreciation), and above these is the profit. If we know in % terms the values of the variable and fixed costs, then we can make a simple break-even calculation for calculating the critical turnover (from or to profit), by the following equation:

$$R_{crit} = 0 = - \text{ fixed costs} + (1 - \% \text{ variable costs of the total costs}) \times \text{Turnover}_{crit}$$

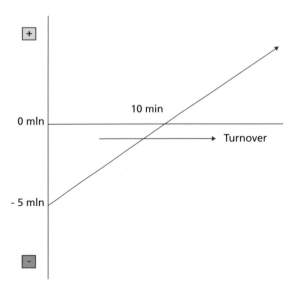

Figure 1.2 Relationship between turnover and costs

Note: These two relatively simple calculations of the ROI and the critical turnover will give an insight into how healthy the company is in terms of being ready to invest in the development of new products.

1.2.2 Life cycle of a product

The next model is the so called **life cycle of a product**, with each product having its own life cycle. Some products have a relatively short life cycle, see the CD player or cassette recorder, whilst others have a continuously improved life cycle, for example the car which founds its form roughly 80 years ago and is still the subject of improvement. Each company with a range of products has to know what is, or can be, the life cycle of these products. The life cycle of a product has distinct phases (see figure 1.3): development, introduction, expansion, stability (maturity), end of life.

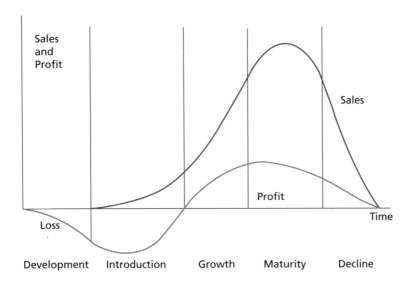

Figure 1.3 The product life cycle

During the **introduction phase** there are mostly losses due to the relatively high costs for market launching. Mostly the buyers are so called ´early adopters´. It is possible that a product will not reach the expansion phase and is taken from the market and thus never attains the profitable stability phase with the highest turnover and profit. The **expansion phase** is the most important phase of the life cycle because it becomes clear if the market accepts the product and what the total potential of a product is in terms of turnover and profit. During the **stability phase** the company has to try to lengthen the life time of a product by continuous improvement activities. Products in this phase are often named ´cash cows´ because they guarantee a stable income for the company over a long period of time. **End of life** products see a continuous decrease in turnover and profit.

Ideally the product portfolio of a company should contain products within each phase, so some newcomers in the introduction phase, some in the expansion phase, hopefully the most in the stability phase and some in the end of life phase (see figure 1.4). It is healthy when this results in a continuum in product development of new products, which can compensate for the loss of the disappearing end of life products.

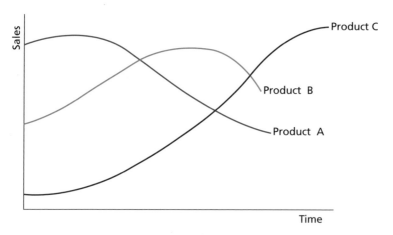

Figure 1.4 A well-balanced product portfolio

1.2.3 Portfolio matrix

The last model that we introduce in this chapter is the **portfolio matrix**, which characterizes the products as: ´stars´, ´cash cows´, ´dogs´ and ´wild cats´ (see figure 1.5). For the company the (rising) *stars* and the *cash cows* are the most important, because they promise income. The *wild cat* is a new comer and is making losses, it has to move to the profitable star position as soon as possible. If this not the case, then it should be stopped as soon as possible. The *dog* is past its profitable life and is already making losses and has to be stopped or sold as soon as possible.

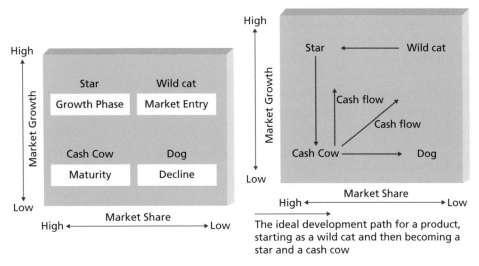

Figure 1.5 The portfolio matrix

The product portfolio of a company and its associated life cycle is a very important factor in the economic value of a company and the control of this portfolio is of great importance and demands the high priority attention of top management, as does the product development process. The IDE process offers a number of possibilities to make this process a successful one, including minimizing the risk of failure.

Note: A portfolio analysis of the product family of a company gives an insight into the potential of the existing products and can be helpful as input for a future product development program by improving existing products and giving directions for new ones.

Example from a study of a well-known electronic form. From 1000 purely new ideas, typically only 100 actually result in launched products and from these there is only one that is successful and becomes a real cash cow! Another 15 may reach a profitable situation and the rest are commercial flops!

Most technical people have to realize that product developments processes are the same as business processes in that their main goal is to make future profits for a long period of time.

1.3 IDE as a Business Process

The main goal of each business process is to create added value for the organization. The opportunities for organizing improvement of the process should also be part of a successful business process. Looking at a business process in general, very often three main areas of attention can be designated. These areas, which can be referred to as the **3 RP Fields for a successfully Business Process**, are:

- the Right Product Field (see chapter 4);
- the Right Process Field (see chapter 5);
- the Right People Field (see chapter 2).

This means that when analyzing or optimizing a business process, these three main areas require attention in terms of: 'are we working with the right things?'. Each area has to be at the correct level in order that the business process can be executed in the optimum way.

Thus, the right product in the right market is not enough for the success or continuation of an organization, it also requires the use of the right manufacturing processes to produce this product and the availability of the right people. Information technology plays a very important role in this business process as a utility (see chapter 3).

A very good manufacturing process applied to a badly designed product will give poor financial results. An excellent product produced by competent people in an organization with a poor financial process can result in a ruinous situation. In this book the IDE process will be threaded as a business process, so it has to be proved what the contribution of this process is in creating an environment for continuously generating and improving added value.

When the IDE process is described, attention has to be paid to the way(s) to develop the right product, which high quality processes have to be applied for designing, engineering, manufacturing and operating this product, and what types of competent people are required to fulfill these tasks successfully?

1.4 What is an Integrated Design and Engineering Process?

Integrated Design and Engineering (IDE) as a business process is defined as a new way of thinking and working about the construction and execution of an integrated development, design and engineering process for products and services, seen over the whole life cycle.

The new way of thinking means an integrated approach during the design process of the wishes of clients, life cycle thinking and system engineering, especially functional thinking, and the application of engineering database(s) for the storage and reuse of proven knowledge.

The 'wishes of clients' means listening to the Voice of the Customer (see also chapter 4) and reflecting all the aspects of these wishes in the design. By thinking over the life cycle during the integrated design process, all phases of the life cycle like: develop, design, engineer, build, install and commission, operate and maintain, and recycle applied materials, come into consideration, including the life cycle costs. Other aspects to consider include: ability to make the product, ergonomics, use of energy, environmental aspects, sustainability, maintainability, and reusability. System engineering means in the first place translating client wishes into functions that must be fulfilled instead of immediately developing solutions. By defining a product design in terms of functions, sub functions, etc. and the possible solutions for these functions in functional decomposition models, it is possible to store in a structural way all the knowledge about this design in engineering databases. It is then possible to continually reuse this knowledge in a very elegant and profitable way.

1.4.1 Characteristics of the IDE approach

The working method of integrated design and engineering is characterized by executing the integrated design process through multidisciplinary teams, who have to utilise concurrent engineering principles. So, they do not work sequentially on the design phases, but rather concurrently. Here all the phases of the lifecycle, together with additional aspects, will be viewed in a balanced way. On the basis of equality, the client wishes are considered in terms of added value before being developed into a product.

For a client or the person commissioning a product, this working method means that their wishes in relation to the design will be executed in a structured way, taking into account all the phases of the expected life cycle, and providing the best possible solution at the lowest life cycle costs. Creating added value as much as possible will be a primary consideration.

Companies that choose this structured way of working not only have very satisfied clients but also for the first time there is the possibility to store their knowledge about functions and products in engineering databases and to reuse this in effective ways, resulting in significant time savings and cost savings through improved and quicker working and fewer mistakes. For these companies there are not only consequences for cost savings but also for their future profit potential, because in general the generation of enough profit is one of the main factors in surviving.

1.4.2 The IDE space

The core or **kernel of Integrated Design and Engineering** can also be explained by the model of the next figure (see figure 1.6), the so called 'IDE space', characterized by the three visible axes and one imaginary.

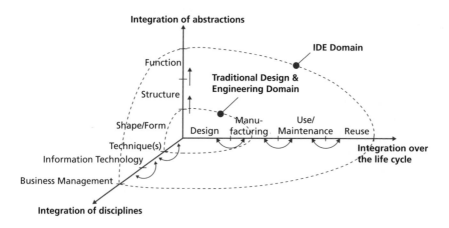

Figure 1.6 The IDE domain or space

- The vertical axis is the '**product axis**' and represents the abstraction of thinking from a system engineering perspective It shows that that in addition to the form of a product, attention must also be paid to the structure and functions on which the design of this product is based. Within IDE the art of abstraction is a necessary competence for a effective control of design data. By applying functional decomposition models for decomposing the product function(s) structure it is possible to store in a structured way in engineering databases the data of the function tree of the product with solutions for each function. An elegant reuse of this data and knowledge is then possible.
- The horizontal axis is the '**process axis**' and represents the integration of all phases and aspects of the life cycle of a product. It shows that during the integrated design process the product life cycle plays a central role in this process. Aspects like marketing and sales, development and design, production and installation, operation and maintenance are all considered.
- The third axis is the '**multidiscipline axis**' and shows the integration of the disciplines during the IDE process, such as technical disciplines, ICT and business (sales and organization). The contribution of the ICT technology is of particular interest for building an engineering database in which all the design data can be stored. The ICT technology makes it possible to share knowledge and to introduce

new forms of cooperation within an organization, over the traditional boundaries of a hierarchic structure.

- A fourth axis is an ´**imaginary axis**´ which represent the cooperation of stakeholders in the IDE process within a multidisciplinary team. The application of ICT technology will dominate their way of working. Mostly these types of teams are very effective through their efficient use of concurrent engineering principles. Social and being a real team player are the essential competences needed for operating in a successful team.

When the general description of the IDE process is compared with the conditions of an ideal business process, the following points of similarity can be noticed:

1. Main Field: the Right Product
   The right product is represented by: the client's wishes are incorporated into functional formulated specifications and all aspects are designed over the life cycle at the lowest possible life cycle costs.
2. Main Field: the Right Process
   The right process is represented by: the process of constructing functional decomposition models, designing over the life cycle using a multidisciplinary team, applying the concurrent engineering principles, using structured ICT models, and being guided by life cycle cost models.
3. Main Field: the Right People
   The right people are represented by: competent people working in different multidisciplinary teams.

In summary, the **IDE principles** can be formulated as follows:

- client demand oriented or demand oriented;
- client wishes formulated on the basis of functionality (functions);
- structure models on the basis of this functionality;
- design and engineering over the life cycle;
- store engineering data in a structured way on the basis of functionality;
- reuse existing knowledge as much as possible;
- multidisciplinary teams using concurrent engineering principles.

1.5 Role of ICT for IDE in Organizations

The use of ICT (Information and Communication Technology) during the integrated design and engineering process over the whole life cycle of a product will have an influence on the total organization and will also create a demand for changes in relation to co operation between departments. ICT requires a process oriented horizontal working organization. This can result in a lot of tensions for the more traditional vertical working hierarchic organizations. It requires the breaking down of walls between departments and exchanging knowledge through sharing and dividing. Each wall or barrier between departments means, in general, a loss of efficiency of 20% through losses in time, in quality of data, in different priorities etc. (see figure 1.7).

Both the IDE process and ICT need a horizontal oriented process approach with no real barriers between people and departments. This means that traditional working vertical oriented hierarchic organizations will have problems in implementing new ICT systems utilising open engineering databases, and in following the ways of working required by the IDE process.

Figure 1.7 Walls cost a lot of money

Case
During an introductory course on IDE principles the participants were asked to characterize their company as: vertical or hierarchic oriented; horizontal or client oriented, or currently turning from vertical to horizontal. The results were as follows:

- 30% vertical oriented (internal);
- 15% horizontal oriented (process and client);
- 55% turning from vertical to horizontal.

1.6   Integrated Design and Engineering in a Company

The scope of the IDE process is to improve existing products and to develop new ones. To explain this scope the **IDE company model** is introduced (see figure 1.8). The main process of each company is the production function which produces the existing product range, the basis of generating the funds for the continuity of the enterprise. The execution plain represents this main process and requires from the IDE an efficient engineering database containing all the manufacturing information related to the product. In addition, small changes in products are developed by means of a continuous flow of improvements. Along with IDE process, other processes like lean manufactory methodology, Six Sigma, etc. are applied, and teams in the form of small groups deal with the continuous flow of improvements.

The development plain represents the development process for new products and major changes to existing products. On this plain the full IDE process will take place with multidisciplinary teams. These teams can be very successful through applying the methodology of concurrent engineering.

The policy plain represents the control processes for both plains beneath, giving the directions and the (financial) facilities for developing new products, as well as improving existing products and the main processes.

The entrepreneur plain (on the rear side) represents not only the control function of the processes of the three plains, but also the information exchange and the processes of continuous learning of the organization in relation to the improvement of processes and products.

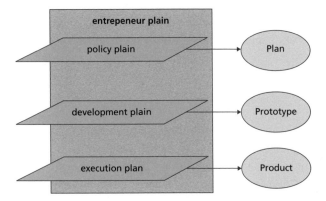

Figure 1.8 The IDE company model

1.6.1 Conditions for applying the IDE process

The application of the IDE process in an organization can often mean a review of the internal processes and even a re-organization. When applying the IDE way of thinking and working, organizations have to fulfill at least some conditions called 'the integration principles'. These integration principles are as follows:

- horizontal client and process oriented;
- clear goals (e.g. for product development);
- organization drivers for innovation and continuous improvement (changes, teams);
- use of ICT technology as a real business process for the storage and reuse of knowledge;
- working with IDE principles;
- personnel who can handle and execute the IDE processes and will, therefore, have the right IDE competences.

Case

A case study is undertaken by companies that already apply (more or less) the IDE philosophy on the above mentioned six conditions. Each condition could be marked with a score of 0 to 100%. The mean outcome per company varies from 60% to 75%. The conclusion has to be that a company must have at least a 65% score to fulfill the conditions to apply the real IDE philosophy. These six conditions and the outcome of their scores were now applied to give a state of art condition measurement of a company, about their capability to introduce the IDE philosophy (new ways of thinking and working)

We will return to many of the above items throughout the course of this book as conditions of applying IDE successfully, including the required competences for the personnel in the company.

1.6.2 The IDE competences

The application of the IDE process requires people with the right qualifications and competences. It is acknowledged that personnel of all ranks must have the right qualifications for doing their jobs properly. However, specific competences are necessary in order to apply the IDE processes (see also chapter 2). Furthermore, applying IDE also requires new jobs, especially stage-managers or directors.

In order to ensure that the IDE process is a successful business process within the organization, IDE directors (just as in a film) are needed to secure the best performance from everybody in the process and ensure that the team will really act as a team (see figure 1.9). It requires more than the commonly known role of project manager. These directors can also act as members of the total IDE process team.

Figure 1.9 IDE means working together

Individual Responsibilities

The modern engineering function is a mixture of technical and economical aspects designed to bring an asset to the highest level of performance for the lowest possible costs over the life time of the asset. Therefore the organization must stimulate co operation between the departments involved with these aspects (design and engineering, production, quality and maintenance departments). It will also require engineers with the following **IDE competences**:

- Technical (a good practical and theoretical level, knowledge of risk analysis techniques);
- Economics (calculations related to the assets, such as: OEE, ROI, PBP, LCC, TCO, etc.);
- Team worker (can work as a team player in small teams);
- Social behavior in relation to the team (also competent in written and spoken communications);
- ICT (3D CAD, CAE, PDM/PLM and CMMS systems) for the information needs of the engineering function;
- Capable of working on basis of concurrent engineering principles.

1.6.3 What will the IDE process bring?

In organizations where the IDE process has been implemented successfully, the following advantages can be put forward:

- can develop better products, that adhere to the client's wishes;
- can manufacture a cheaper product, often 30% cheaper than expected;
- have better internal processes, faster and without failures;
- have a better insight into manufacturing costs, and also into the contribution made to the profit of a product;
- can achieve profitability through the application of ICT;
- can earn a lot of money by delivering optimal long life service through better handbooks, services and parts;
- have the right people to execute these processes.

The right integration of processes and people in these organizations will give them a lot of flexibility and scope for future developments in relation to markets and products.

1.6.4 Where is IDE already being applied?

IDE as possible way of working is already applied in the fields of aeroplane and car manufacturing, in the field of consumer electronics such as mobile phones, etc. and in high value machine equipment. Some new and interesting fields are:

- *High tech mechatronics:* Nowadays the world of machine equipment builders is rapidly changing. From pure mechanic oriented, via electronic controls and PLC controls, to fully computer controlled. So a lot of machine equipment is now controlled by computers and embedded software. See the modern machine centre for cutting, drilling and turning of metals in one machine. Another example is the wafer stepper for making chip patrons on basis of lithography. Without the help of IDE these developments would be slower and less profitable. Development teams have to work on the basis of concurrent engineering.
- *Building environment:* New buildings like houses, offices, hospitals, etc. have to be built with sustainability in mind - with low energy consumption, zero CO_2 emission, high standard of comfort, low life cycle costs. Also building processes incorporating design, construction, operation and maintenance in one project must be done by building consortia with a lot of IDE competences in their group of companies. Examples are the modern tunnel complexes with a lot of electronic control systems, which have to work in an integrated way. In the more traditional way of constructing buildings, with a high domination of traditional architects (purely focussed on ´form´), builders only focussed on lowest prices, and with poor cooperation between the building services sub-contractors, the realization of this type of building is very difficult to achieve. Only the IDE way of working can create the right circumstances to do this. An example is the creation of the ´city of the sun´ (Heerhugowaard, the Netherlands) where a big city plan is under development to create houses, schools,

offices, shopping centre, etc. with as little energy consumption as possible, utilizing the sun energy as much as possible, and with the lowest CO_2 emission (see figure 1.10).

- *Industrial maintenance on high value equipment*: The IDE way of working can also be very successful for organizing industrial maintenance on high value equipment, giving effective production rates of high volume at high quality. Here the integration of production, maintenance, quality and engineering has to take place in order to achieve the highest (quality) output against the lowest operational costs, or the highest asset utilization factor.

Figure 1.10 City of the sun [Balloon Picture © Maarten Min, Min2]

Summary

In this chapter we have introduced the concept of the Integrated Design and Engineering process as a booster for product development processes. We have explained in general terms that product development has to be seen as purely a business process because it is too risky to see it as an isolated technical issue. The methodology can be applied for both process improvement and development plans for products.

We have also explained that in order to apply this IDE process in an organization, the organization has to open the walls between departments and the people in the organization must have the right IDE competences to make this way of working successful. The results of this way of working, with a high level of integration between the activities, are typically very profitable for all the stakeholders involved in product and process development projects.

Exercises

1. What are the pro's and the con's of introducing and applying the IDE process methodology in an organization?
2. Which competences can be specified as IDE competences?
3. Is your organization ready to apply the IDE philosophy? Take the six conditions (see § 1.6.1) and give each condition a mark between 0 to 100%. Calculate a mean value of the outcomes of each condition.
4. If the income/turnover is €375 million/year, the costs are €300 million/year, then the cash flow is € .... million/year. The total value of investments is €300 million, the depreciation is 10% of total value of investment. What then is the profit and the ROI?
5. If fixed costs are €25 million/year and % variable costs are 40% or 0.4, what is the critical Turnover$_{crit}$? What is the meaning of this Turnover$_{crit}$ for a company?
6. You are the owner of a traditional machine manufacturer and are thinking over the future. What could your innovation agenda be through the application of the extended simple business model?

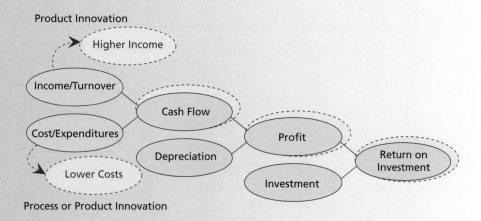

Product Innovation

Literature

- Shah, J.J. et al., *Research Opportunities in Engineering Design*, Final Report NSF Strategic Planning Workshop, Arizona State University, Phoenix, U.S.A.
- Clausing, D., *Total Quality Development , A Step by Step Guide to Word Class Concurrent Engineering*, ASME Press, 1998, ISBN 0791800695.
- Grant, R.M., *Contemporary Strategy Analysis*, Blackwell Publishing, 2002, ISBN 0631231366.
- Roberts, J., *The Modern Firm*, Oxford University Press, 2004, ISBN 9780198293767.
- Ulrich, K.T. et al., *Product Design and Development*, McGraw-Hill, 2004, ISBN 0072471468.

Chapter 2
IDE Philosophy Demands for Team Working by Concurrent Engineering

2.1   Introduction

The development team always plays a crucial role in the successful development of a new product or process. The composition of this team is of great importance, therefore we will take a lot of care in ensuring that the composition of the team is right. When this team applies the IDE philosophy and has the right IDE competences, than there is a good chance that the team will finalize the development process on time, with the right product and the right production process. An overview of the appropriate IDE competences is given in this chapter.

The starting points of concurrent engineering principles for team work are introduced as the way to organize and plan the development work. Also the Lean principles are applied for the development process.

To make sure that the team can work successfully, it is not only the team composition that is important but also the organization of the company, which has to fulfill certain conditions. A horizontal oriented process organization with open borders between departments can often offer these required conditions.

2.2   What is the Right Team?

Composing a development team in line with the IDE philosophy is a process that has to be done very carefully. A lot of aspects have to be fulfilled. The **IDE team** has to be multidisciplinary, have knowledge about design and engineering aspects, have IDE competences, be able to act as a real team and also to apply the concurrent engineering principles.

Workers in a team will be become team players which means that they play a team role.

R.M. Belbin, who has done a lot of research about working in teams and team roles, has defined the nine well known roles found within a team (for problem solving in a team). Successful teams consist of a fair mix of these roles. Every person has mostly one strong role and two or three other weaker roles. The **Belbin problem solving team** roles are:

Action oriented by:
1 Shaper
Strong: The shaper is the dynamic partner in the team. Has great drive, is dynamic and defiant. Can work very well under pressure.
Weak: Can provoke too much in an aggressive way.
2 Implementer
Strong: The implementer changes ideas through practical solutions or actions. Has discipline and is secure. Is reliable, is conservative and efficient.
Weak: Can be inflexible and conservative, and slow in introducing new possibilities.
3 Completer-finisher
Strong: Finds the mistakes of the others or things that are forgotten, and delivers on time. Is conscientious and works with precision and care.
Weak: Worries too much about the progress of a process. Can not delegate.

Human aspects oriented by:
4 Coordinator
Strong: The coordinator is a human oriented leader. Clarifies the goals of the team, guides the problem solving process and can delegate tasks.
Weak: Can be a manipulator or not make decisions.
5 Team worker
Strong: Is the lubricator in the team. Listens, builds relations, prevents friction to enable the team to work better. Has care for cooperation in a team, is mild and diplomatic.
Weak: Does not like disagreement so is mostly a subjective decision maker.
6 Resource-investigator
Strong: Researches opportunities for ideas, is extrovert and communicative. Is good at nurturing new ideas for further investigation.
Weak: Can lose interest after first wave of enthusiasm, can be over enthusiastic.

Problem solving oriented by:
7 Creator or Plant
Strong: Solves the most difficult problems, is a specialist in generating new ideas. Has

a high IQ and is more introvert. Is good at creativity, fantasy and out of box thinking (unorthodox).

Weak: Is ineffective in communication, and does not like practical details.

8 Monitor-evaluator

Strong: Has an overview of all possible options (big picture). Good at focussing on an objective decision process with the right outcomes, is strategic, sober and attentive.

Weak: Has a low drive and is not inspirational to others.

9 Specialist

Strong: Has unique knowledge and experience. Is loyal, purposeful and self started.

Weak: Contributes only on a relatively small area and can be lost in details. No interest in other discipline's troubles or ideas.

Through the use of a **self perception matrix** everybody can make a profile of themselves (see figure 2.1). Mostly one of the nine roles is dominant and the other roles are more or less present. No-one will play exactly the same role in each team, because everybody has their own mix and other people bring in their mix. In a radar chart (shown in figure 2.1) we can give an impression of each individual role. When the results of all team members are placed in a radar chart then the behavior of the team can be predicted and if the result gives a one-sided picture, then compensating measures have to be taken in order to create a balanced team.

On the other hand, the team has to be a multidisciplinary product development team, with design and engineering capabilities as the main competence. D.G. Ullman indicates that the following disciplines can be found in a multidisciplinary development team:

- product design engineer;
- product manager;
- detailer;
- drafter;
- technician;
- materials specialist;
- quality controller;
- industrial designer;
- assembly manager;
- vendors or supplier representative;

Note: Missing disciplines in the enumeration of Ullman are the operational and maintenance roles.

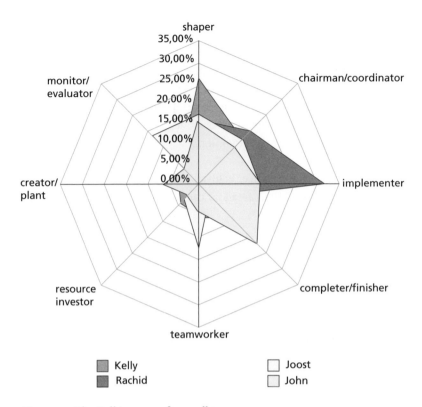

Figure 2.1 The Belbin score of a small team

Mostly a good working team consists of 6 to 8 people, so not every discipline has to be a constantly available in the team; depending upon the stage of the product development process, the composition of the team can change.

A third point of consideration for the team members is knowledge of the IDE philosophy and the associated IDE competences. These specific IDE competences can be summarized as follows:

- can handle the principles of system engineering, especially the formulation of client wishes into functions and functional specifications;
- can apply the principles of Life Cycle Engineering, which includes all the phases and the different aspects of the life cycle, also including the life cycle costs;
- can make functional decompositions of the function that has been defined by the client's wishes;
- is familiar with the common applications of ICT related to IDE and is comfortable working with these applications;

- will have some market experience in designing and producing saleable products;
- can work as a team player in integrated design teams;
- has good social behavior in relation to team (group) dynamics.

2.2.1 IDE team leaders or directors

We have already mentioned that the leader is of great importance in managing an integrated development team, and a director role is therefore also very important. In addition, it is possible that over the whole of the product development period or time the specific content of a director's role will change. Directors play an important role in making the members of the team real team players. Specific IDE director roles can be:

- the product definition director, who will lead the team in defining the right functional specifications (derived from client wishes);
- the product development director, who will lead the product development process over the life cycle;
- the product support director, who will lead the after-sales services and also bring in continuous improvement ideas from clients of existing products;
- the product structure director, who will optimize the process of product structures in relation to the ICT applications and the reuse of knowledge;
- the human resources director, who will be involved in continuously improving the personnel skills and social competences.

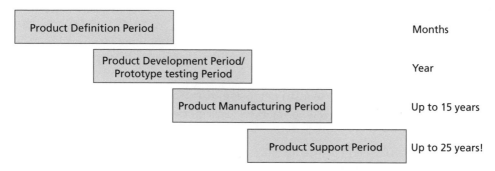

Figure 2.2 Time frames for product stages

When we represent the time frames for the stages of a product (see figure 2.2) we will see that the development of a new product is initially dominated by product definition work, there is then a period with design and engineering work, the manufacturing stage and then the installation and operating period or product support period. Note that the periods of manufacturing and product support can take many years.

Each period has its own priorities, so the team composition of disciplines can change, but at each stage the team members have to work together using the methodology of concurrent engineering.

2.2.2 Personal characteristics of a team member

Each team member brings their own personality with professional, cognitive and social skills. Belbin describes the possible roles in teams, but each person brings in a unique combination and has to fit in with the team for both personal and team success. Central to each team is the level of problem solving capability. If this capability is too low, the team can fail to find the right solution for the problems it faces.

So the problem solving capability of each team member affects how problems are solved and has a strong influence on the team. **Myers and Briggs** developed a model **(Type Indicator)** with four dimensions that describes how individual people deal with problems in order to solve them. The four dimensions of individual problem solving styles are described below:

Introvert-extrovert
An introvert person is a good listener, is reflective, thinks and then speaks, and enjoys solving a problem alone.

An extrovert person is full of visual energy, has a lot of interactions with others, tends to speak before thinking.

Most individuals are sometimes more introvert and sometimes more extrovert, depending on the situation.

For a team, the extroverts have to learn not to dominate discussions, whilst the introverts have to be encouraged to bring their ideas more confidently to the table.

Facts-possibilities
A fact person deals with facts and details, is practical and realistic, and appreciates the here and now. They need encouraging to fantasize, to think more expansively and allow others to do the same.

A possibility person thinks in terms of possibilities, concepts and theories. They are looking for relationships between pieces of information and the content of the information. They need encouraging to focus more upon details, specifications and the main issues.

The product development process requires working with both facts and possibilities. So both types of thinking are essential for a successful development team. Facts versus possibilities can be the cause of a lot of miscommunication, misunderstanding and team problems.

Objective-subjective
An objective person is logical, detached and analytical. They have an objective approach to making decisions. They have to learn to deal with incomplete and uncertain information.

A subjective person makes decisions based on interpersonal involvement, circumstances and the right thing to do. They have a subjective approach to decision making.

Usually engineers are trained to make decisions based on objective measures, but in practice the information is often far from complete and inconsistent in its quality, so it requires a subjective approach. The design team has to be open to a variety of approaches to information collection and to a range of decision making styles.

Decisive-flexible
A decisive person seems to make decisions with a minimum of stress, and prefers an environment that is ordered, scheduled, controlled and deliberate. They have the tendency to make decisions by jumping to conclusions, so need to learn to consider alternative solutions.

A flexible person goes with the flow, is flexible, spontaneous and adaptive, but finds making decisions and sticking with them difficult. They have the tendency to not make a decision at all.

For a design process the decisive approach can be too strong because only one possibility is evaluated, whilst the flexible approach can be too weak because the decision process is too slow.

Note: Different aspects of these four areas of a person's problem solving behavior have already been covered in the nine team roles by Belbin. So this more personal approach to describing behavior can be very helpful in building the right team.

2.3   Building the Product Development Team

We have seen that there are a large number of considerations to take into account when building an IDE product development team. In the first place the disciplines required over the full life cycle, then the personal behavior in relation to problem solving, further the IDE competences, and finally the right mix of team roles. We have seen that a lot of personal problem solving capabilities are also in the team roles and, for the IDE competences, we can state that in order to be a successful IDE team these competences have to be available over a number of team members. Team members should not have only technical knowledge but also the right social and behavioral skills.

So, in order to check that the team has the right composition we can use the following matrix, with the disciplines on the one axis and the team roles on the other axis (see figure 2.3.)

Discipline/Belbin	1	2	3	4	5	6	7	8	9
Product manager				X					
Design engineer		X							
Detailer			X						
Drafter						X			
Technician								X	
Materials specialist									X
Quality control	X								
Industrial designer							X		
Assembly manager					X				
Vendor / supplier			X						
Product support							X		
Client representative									

Figure 2.3 Checking the composition of the team

In this matrix each row and column should have at least one mark, so that all the Belbin roles are more or less fulfilled. A suggested team composition is shown in the matrix.

Teams also have specific requirements, because the team members have to cooperate as a team. These requirements are:

- team members must learn how to collaborate;
- teams have to make decisions, so team members must compromise to reach them. Consensus leads mostly to more robust decisions;
- team members must communicate in order to achieve real-time problem solving;
- team members must be committed to the team goals.

2.3.1 Coaching

Because working in a team has a great impact on the social capabilities of the members, a lot of attention must be paid to the individual team members, since their social behavior is one of the success factors of a team. A good inauguration meeting in a tension free setting, (some) training in team social behavior and a team coach for team building is of essential importance for team success. Every team member has to understand the impact of team dynamics on personal behavior, on stress, on conflicts, on problem solving and on their own character. Also there must be continuous work on building the team performance in aspects such as working on clear and realistic goals, the rules of team behavior, and spending time together.

2.3.2 Teams over the life cycle

The composition of the product development team changes over the life cycle, because during the development of a new product the focus on specific requirements will change. When investigating or formulating product ideas, the wishes of clients are the drivers, so product definition will become the main focus and the team leader or director will be a person from that discipline. The next phase is product design, and engineering is the main issue so the team leader will come from that discipline; this is followed by product manufacturing and product delivery with product support as the main issues. Each phase will have its own specific main focus and associated team composition. It has to be very clear that all important aspects of a new product must be represented in the team, in some phase acting as a representative providing input, and in the other phase being a member working on solutions. E.g. representatives of production and product support (maintenance) have to be team members from the beginning.

2.4   Concurrent Engineering

The product development team has to work over a long period of time, covering the whole of product development, from idea / concept through to the operational product in full service. In order to work faster and more efficiently, the team members representing the different disciplines will carry out their work more or less concurrently. This means that when the product idea definition comes to a certain stage, the work on design and engineering, on production methodologies and product support (operation and maintenance) will have already started, of course with some delay in each discipline. Thus, there is no waiting until one aspect or discipline is completed and delivered (see figure 2.4), but starting as soon as possible, concurrently with each other.

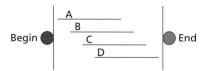

Figure 2.4a Working together through CEP (Concurrent Engineering Principles)

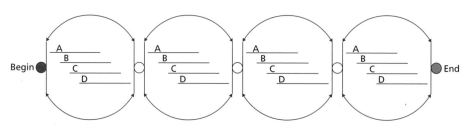

Figure 2.4b CEP Peristaltic Planning Model

Special attention has to be paid to the planning of those CEP activities which have to be undertaken concurrently. When working on a project, e.g. the product development project, which may take some months up to a year, it is helpful to define some points or stages at which decisions are made and mark them at the end of each phase, so everybody is together on the same point in the process.

The big advantage to working inline with concurrent engineering principles is that, in a relatively short period of time, all the stakeholders of the process have to work very closely together with regard to all of the aspects of the integrated design and engineering process. All the stakeholders will start as soon as possible with their activities and are also involved from the beginning with the main choices and decisions in the design process.

2.5   Lean Development Process for Product Development

The development of a product can be implemented as a business process following the Lean methodology for process improvement and optimizing. **The Lean Process Methodology** contains certain phases and steps for improving business processes. We will follow these phases and steps and then apply this methodology for the Integrated Product Development Process.

The scenario to Lean consists of the following six steps:

- Phase 1: Formulate. What is the mission and the vision of an organization?
- Phase 2: Define the Critical Success Factors.
- Phase 3: Define the Performance Indicators.
- Phase 4: From ' Ist' (current state) to ' Soll' (future state) in a process.
- Phase 5: Structural Improvement by Lean (seven steps plan).
- Phase 6: Continuous Improvement.

The Lean process itself consists of the following seven steps:

Step 1: Start the Preparation by:
- Organization of team(s);
- Formulate what is right and what is wrong and what you are doing to correct 'wrong';
- Create order and tidiness;
- Define the Performance Indicators.

Step 2: Make a chain of the processes:
- Organize flow in the chain of the processes;
- Make a chain of the processes.

Step 3: Work as defined and collect data:
- Work according the defined flow of processes;
- Implement the so called order and tidiness by:
    - Sort;
    - Structure;
    - Clean;
    - Standardize;
    - Maintain.
- Collect data of processes.

Step 4: Improve:
- Use data for Pareto analysis, analyse cause and effect;
- Make plans for improvements and execute them;
- Make graphs to follow the trends.

Step 5: Evaluate and change, if needed,the processes:
- Optimize the flow of processes, improve lay outs;
- Do small investments to support the improvements;
- If speed of improvement decreases: time for significant change.

Step 6: Make improvement a continuous process:
- Continue improvement, data collecting, 5S adaptions, improvement management;
- Study new methodologies for improvement of processes, time studies;
- Standardize by fixing improvements in work flow documents.

Step 7: Dot the i's:
- Evaluate the cooperation and team work of the team(s);
- Improve, if necessary, the team organization;
- Formulate sharper goals and continue the improvement activities.

We will now look at how to apply the Lean principles for the product development process in order to improve the way in which this process works.

Step 1: Start the Preparation
Organization of the design or product development team (see Belbin methodology). Typical points for improvements can be:
- Lowering of the throughput time of the whole process;
- 'Just in time' delivery of drawings and other information;
- Increased output of drawings per week;
- Development of prototypes 'in first time right', so reducing losses of money and time.

Create order and tidiness by:
- Register the appointments;
- Good and structured stored information of all kind;
- A well organized information data bank containing all the essential information, and fast access to re use existing knowledge;
- Launching a structured machine and tooling library.

Define the Performance Indicators work in time, production and quality.

Step 2: Make a chain of the processes
- See Product Development Process (Chapter 5);
- Ensure the flow of the work process conforms with the defined stages, each stage or phase will be finalized by a document containing all decisions and outcomes of the process;
- See Concurrent Engineering (section 2.4);
- Product development teams have to work according to the concurrent engineering principles.

Step 3: Work as defined and collect data
The real work flow done by the product development team must be guided by structure models and structured code systems. These important modules and structures are introduced in Chapter 3. The accessibility of data and information is crucial, so that the re use of this knowledge can be applied.

Step 4: Improve. No real case, not applicable here.
Step 5: Evaluate. No real case, not applicable here.
Step 6: Make improvements. No real case, not applicable here.
Step 7: Dot the i´s. No real case, not applicable here.

2.5.1 Impact of Lean upon the organization

Working with Lean principles for business processes has an impact upon the organization. One of the great consequences is the so called 'turning' of the organization vision from 'top down' to 'bottom up'. Within a successful organization the executive part which sets the rules for the business processes is the most important. There the products are produced and the income generated. The management and the staff have to support these executive processes and improve them continuously. Not supporting these processes can result in these staff departments producing 'waste' which can ultimately result in these departments being made redundant.

Lean means working with multidisciplinary teams and offers lot of possibilities to organize and implement integrated improvements.

Another important factor is the way of thinking about organization building. IDE (and modern ICT and Lean methodology) asks for horizontal oriented process thinking with open borders between departments (removing the walls), which conflicts with the common thinking of vertical oriented hierarchies. When an organization is not able to make the move from a vertical orientation to a horizontal one and is not able to remove the walls between departments, then it is also not able to introduce IDE (and Lean and other improvement processes) because the importance of a department is greater than the importance of the company as a whole!

2.6   Team Work in the Building Environment

Team work in the building environment takes place at least at two levels. First at company level, we see that for the design, building and construction, operation and maintenance projects, companies are forming partnerships to work on a collaborative basis on these projects, often also applying the concurrent engineering principles. All of the work undertaken within such a partnership has to be done on the basis of IDE. Also in all of the companies involved, there will need to be cooperation between departments and members of the project teams on the basis of concurrent engineering. A pyramid of teams is often composed for building processes. On top the main team that is responsible for client contacts, coordination, planning and budgeting. There are a number of project teams controlled by the main team, and finally there are the internal company teams, again often controlled by project teams (see figure 2.5)

Figure 2.5 Teams in the building environment

As an example, a simple house (see figure 2.6) with an integrated heat pump system as comfortable heating system that has the heat source in the ground of the garden, the heat pump in the cellar and the heat delivery in the floors and the walls. Such a system can only be successfully designed and constructed by excellent co operation between the designers and the makers in the team of companies involved.

2.7   Team Work and Education

In modern education group work or project work, small teams (4 to 6 persons) are now used. The problem for teachers, who are specialists, is to work effectively with groups when there is often a lack of knowledge about group dynamics, so that group processes are frequently not really managed. To overcome such problems, an introduction to small group education is included below.

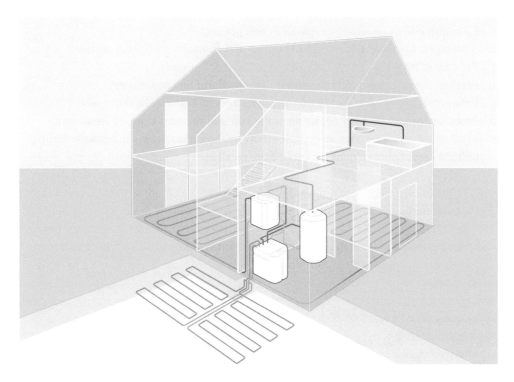

Figure 2.6 Modern heating system [© Stiebel Eltron b.v.b.a., Brussel]

We will start with IDE competences and education. We have seen that the most important IDE competences are:

1. Can think in terms of functions in relation to client wishes, and can formulate these wishes into functional specifications.
2. Can make functional decompositions with the functions defined from client wishes.
3. Can apply the whole life cycle as a basis for the design and engineering process, including cost.
4. Can make ICT structure models (including codes) from functional decompositions.
5. Can think as a business person with an understanding of entrepreneurship, market feeling and marketing.
6. Can work together effectively with other disciplines in the team, applying concurrent engineering.
7. Has appropriate social behavior in relation to team (group) dynamics.

When we look at a manufacturing and engineering firm, we can see that, in general, the professional functions around the design and engineering departments are (see also figure 2.7):

1. Entrepreneur, as booster for new ideas and products.
2. Marketer and salesman, as a client contact and formulator of client wishes.
3. Product manager, as formulator of the product definition.
4. Product designer, as maker of the first design (form and prototypes).
5. Product engineer, as maker of all the product specifications 'how to make it'.
6. Process engineer, as representative of the fabrication and testing of the product.
7. Product support engineer, as representative of the installation, the commissioning and the operation and maintenance of the product.

Figure 2.7 Professional functions in design and engineering

When formulating projects for education executed by small teams of 4 to 6 persons, ensure that each person plays a certain professional role or function, bearing in mind the IDE competences. Formulate these roles in advance of starting the project in combination with the main goals of the project.

The small team must, in addition to concentrating upon the technical solutions of the project, also report upon the team effort and the outcomes of each team role. Every team member should also report upon their own experiences of playing their role in the team. This means that the tutors, who are guiding these small teams, have to assess both parts of the project. This can result in a common mark for the team report and a personal individual mark for the team role, so that everybody can gain a different total mark for a project.

2.8   Management Paradox (1)

When we look at the time that top management spend on putting together product development programs and processes, including the composition of (multidisciplinary) teams (less than 5% of the time) compared with the time spent on routine production activities (more than 60% of the time) then we can speak of a management paradox (see figure 2.8).

This is because during the periods of product definition and design, engineering and all the parameters of manufacturing, the cost price of a product and hence the future profit of the company is decided. But during these very important periods, the key decisions are made by teams with little or no involvement of top management. Sometimes the situation is even worse, with the work of the design and engineering departments separated from that of production and after-sales, so that the influence of these departments on the outcomes of the product development process is negligible.

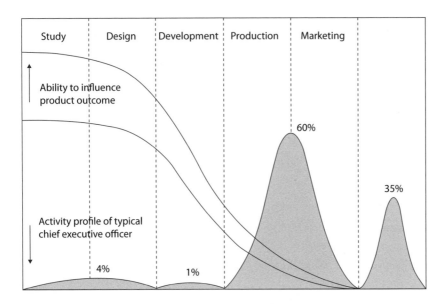

Figure 2.8 The typical role of the CEO in new product programs

The preferred situation is that top management must get involved in the early stages, take care in building the right teams, and pay attention to stimulating and supporting cooperation between all of the stakeholders. In addition, clear goals for the outcomes of the product development process have to be defined.

Note: During the initial product development stages (shown as 'study' and 'design' in the figure) the possibility of influencing the product outcomes is the greatest, whilst the time spent on these activities by top management is the lowest.

Summary

Following the IDE philosophy, the composition of a team is the most critical factor for applying and implementing this way of working. In this chapter we have introduced the Belbin methodology as a guideline to composing a multidisciplinary team in which the personal role of everybody as team member has to be taken into account. Furthermore it is possible to apply the Lean philosophy as a process way of working for the integrated product development process. Also we have mentioned the desire to change the way of thinking within organizations from vertical internal hierarchic oriented to horizontal client process oriented.

Exercises

1. You are a teacher of a school of engineering and are involved in project work undertaken by small groups. For the new project you will put down on paper some ideas for the team composition. What can be your criteria?
2. You are a team leader in a design office and have to put together a new small team for a product improvement project. You know about the ideas of Belbin. What are your criteria for selecting the team members?
3. How can you organize a team for product development over the whole Life Cycle?
4. See exercise 3. Do the same for the building environment.

Literature

- Belbin, R.M., *Management Teams, Why They Succeed or Fail*, Elsevier-Butterworth-Heidemann, 2002, ISBN 0750626763.
- Belbin, R.M., *Team Roles at Work*, Butterworth-Heidemann, 1993, ISBN 0750626755.
- Clausing, D., *Total Quality Development, A Step by Step Guide to Word Class Concurrent Engineering*, ASME Press, 1998, ISBN 0791800695.
- Ullman, D.G., *The Mechanical Design Process*, The MacGraw-Hill Co, 1997, ISBN 0071155767.
- Groot, M. de. et al., *Small Group Activity*, FullFact, 2007, ISBN 9789078210030.
- Wijnands, H. and H. van den Boom, Scenario to Lean, 2008, Nelissen, ISBN 9789024417346

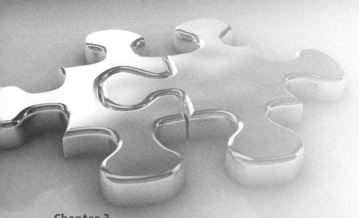

Chapter 3
IDE Structure Models for Product Data Management and Reuse of Knowledge

3.1 Introduction

In the Integrated Design and Engineering philosophy the integration of information is also a main issue. Storing all types of data in a structured way in an engineering database is essential. But historically in many companies there is more than one computer information system still in use. In addition to the mainframe there are, for example, networks and PC systems in operation. The real problem is that each system tends to have its own (code) model for collecting and storing data and, therefore, these systems cannot exchange data in a proper way. Sometimes an asset is stored in three or four database systems under different codes and even under another name. On the other hand, the data of design processes has to be organized so that relatively small parts can be identified. This is necessary to make it possible to analyze design and failure modes at a detailed level. Also most engineering databases are organized as stand-alone systems with no real connections to the main systems.

For bookkeeping and administrative departments the data can be collected at a higher level e.g. production line or product development level. This difference in interest often gives rise to tensions between the information department and the design organization. For the design management it is very important to realize that their interest in information has to be more specific than the higher 'level' wishes of bookkeeping and even production! So design management have to ensure that it is possible for the system to support their interest in specifying a deep enough level in the design process that allows it to be controlled and managed. This means, in practice, three to four levels deeper than the finance and production needs! A lot of ERP troubles arising out of implementing a new system have their origin in failing to meet the wishes of how deep the information models have to go, or in other words, how far the decomposition has to go. The result of this can be the loss of a lot of historic data when it comes to changing systems.

How can we overcome these problems? The only way is for design managers to realize themselves that in order to control design processes through information systems a very proactive attitude is needed. The information database and systems are empty boxes and the structure in these boxes has to be organized around the products and processes themselves, by structured code systems, organized by functional decomposition models. Because products and production processes have a great variety and number of differences in their set up, the outcome of these decomposition models is always more or less 'tailor made'. Furthermore, these products and processes have to be 'cleaned up' in relation to effectiveness and efficiency aspects before the information system can be developed and implemented. Too often existing processes are incorporated into information systems with a lot of unexpected results or outcomes as result.

We will present some structure models that are able to set up the lay-out of such a engineering database. The big advantages of such a database are the access to failure free data and the re use of proven knowledge. Furthermore we will describe a general approach and give some successful examples.

3.2   Information Systems for Organizations

A lot of users of business ICT systems like ERP, 3D CAD PDM, PLM and CMMS systems, are not satisfied with the results of the system and very often only 20% of the potential of the system is actually utilised. Why is this the case? A big problem in a lot companies is the fact that assets and data of assets and processes are registered in different data systems by different codes and even by different names! The internal exchange of information is often impossible. As a result, the systems of bookkeeping, assets inventory, safety and environment and maintenance are handled by different information systems each with their own code systems.

One of the main reasons for this situation, we have discovered, is that most industrial organizations have not paid enough attention to the standardization of the information systems. Even on one site, departments can use different standards. Unfortunately, the importance of a good process and product standard code system is often not recognized. Inherent to the fact that a code system is not seen as primary science, is that there is hardly any literature that supports this.

The selection of a company structure and code (numbering) system is a **strategic decision**, and is one of the basic and key factors required to make an ICT system work successfully. Very often companies are the result of mergers and this can make the situation even worse. This is because every 'old company' has their own system and there is often no drive for standardization. So one of the possible benefits of the merger i.e. a reduction in costs, is not achieved because of the missing code links between the information systems (which means that they cannot communicate with each other).

Another missing link in an industrial organization are people (or a small department), who are involved in setting up, maintaining and controlling the structure and code system. Every design and engineering organization or department should have such a function.

3.2.1 Possible causes for implementation failures

When implementing a new ICT system in an organization, a few factors that have a direct impact upon success or failure arise time and again. The first is that all aspects of the business processes are represented on basis of equality. How are the task forces composed, what was the influence of the work floor, what should be done with the informal information at each level of the organization (the informal Excel sheets)? If the hierarchy thinks that the existing organization will be or can automated, then the outcomes of the process of implementation may deliver a lot of surprises and disappointments. In reality an organization knows its formal and informal patterns, and has a lot of isles with their own specific information needs (the well known private Excel sheets). Deciding how to implement an ICT system without taking in account the needs and influence of these Excel sheets can give rise to a lot of trouble too. Each ICT system is in essence an empty box and is also a horizontal oriented process, which has to be filled with its own business processes and all the relevant data. So when existing processes are automated without a cleaning of these processes on basis of effectiveness and efficiency, without clear goals and performance indicators, and with no recognition of the wishes of all stakeholders, then the results of implementation can be dramatic.

Another important item is standardizing the names, code structures, processes, etc. If it is only possible to organize these items on a department level rather than at an organizational level then a lot of potential advantages are not achieved and troubles are again introduced. Clear structure models have to be available, or to be developed, for processes and products.

Another problem can be the fact that informal rumors suggest that the total cost for fully implementing a system can be 10 times the system costs. Also the input of data or transformation of data from old systems can be potential sources of trouble.

Finally, everybody should be properly trained for working with the new system, but with the high costs in mind, many organizations may try to reduce the costs of implementation by cutting these costs. As a result, if not everybody is trained properly then these persons will not understand the potential of the system as far as their own work is concerned, so it will never be their baby!

Summarized below are the possible causes of failures when implementing large integrated information systems like ERP, PDM, PLM, 3D CAD, CMMS (Maintenance), etc.:

- the task force for selecting and implementing the system does not reflect all the stakeholders in an organization;
- information systems are process oriented and can give rise to a lot of frictions in a hierarchic vertical oriented organisation. Both formal and informal roles within the organization need to be taken into account;
- no transparency in relation to the goals and means of processes;
- structures to store data and information relating to products and processes are not available, or not to an appropriate level of deepness;
- ideas and wishes of the work floor are neglected, so that the system will not be their baby;
- training at all levels is not implemented, especially the lower level in the organization, so that not everybody knows or understands the benefits of the system;
- the new information system can freeze the existing organization (processes) and so introduce a lot of inflexibility. This is especially the case for un-optimized processes (goals, effectiveness, efficiency, performance indicators) where this situation can even threaten the continuity of a company.

3.2.2 What is a better approach?

Organize the implementation process by setting up a task force that exists under the control of a committee and some working groups which represent all the main and support processes on the basis of equality. Furthermore, for every process, there should be formulated the goals, the critical performance indicators, the way of working, the planning , etc. as the beginning of the process of cleaning, which means we are doing the right things effectively and efficiently.

ICT systems are, in principal, empty boxes, so a lot of attention has to be paid to the standard of the working methods, standard of codes, and to product structure models and decomposition models in order to bring these processes and products properly into database structures.

In summary, the preferred approach to implementing an ICT system will be:

- the task force is a real reflection of the organization with all the stakeholders represented;

- the organization formulates what the new ICT system has to deliver, including possible new ways of working;
- all the processes are evaluated in terms of effectiveness and efficiency, and real goals and critical performance indicators are formulated;
- the needs and wishes for the new ICT system are formulated, potential suppliers are selected, and demonstrations are carried out using real facts and figures;
- if possible, consult other users of the proposed ICT system;
- run trials with the system involving all stakeholders, and use their comments for improvements;
- investigate the consequences for the organization of implementing the new ICT system and communicate these to all the stakeholders;
- set up a proper training program for all users and stakeholders;
- set up a department to control structures, standards and code systems for the new ICT system.

Note: It is said that the satisfaction of the lowest level of computer user is the key indicator for the successful implementation of an ICT system.

3.3   Structure Models for IDE Applications

Many sources of information such as historical information of products, processes and process equipment are available within a company. Historical information is the most important information source for any plans to set up a new product or manufacturing plan and essential as input.

Historical data is only valuable if this data is presented as:

- clearly structured;
- contains correct and essential parameters;
- well defined and understandable;
- possibly contains pictures.

In addition to this, two other problems related to code systems are identified and need to be addressed:

- define a code system that allows tagging of all equipment;
- specify the levels of registration in relation to the ERP, PDM, CAD and CMMS system.

Information should be easy to select or find. In order to obtain this, the following aspects should be recognized as typical for a good code numbering system:

- code systems are based on the organization structure;
- code systems are the basis for the information needs;
- code systems are essential for good internal communication.

In organizations we can recognize structure models for analyzing processes and modelling products. For analyzing systems it is advisable to use decompositions models like:

- the IDEF-0 model for analyzing processes;
- the function-product model (the so called Hamburger model) for products;
- the V-model of systems engineering for products.

3.3.1 IDEF-0 model

A methodology that can be applied for making decompositions is the **IDEF-0 methodology**. IDEF stands for Integration Definition Function Modeling (IDEF). This methodology has been developed especially to analyze processes and is normalized for its process levels and codes (code system). A0 is the zero main level for describing the main process. On the first level the sub processes become the codes: A1, A2, A3, A4, etc. On the second level: A41, A42, A43, etc. and on the third level: A421, A422, A423, etc. Of course the decomposition process can go further until it reaches a level where no processes can be decomposed.

From the black box of the main process (this being level 0 or A0), we can find by opening this box the underlying sub processes (level 1) and under these sub processes the sub sub processes (level 1.1), etc. Processes are written as verbs (to make a function explicit, like: to design for the design process, to produce for the production process, to book keep for the bookkeeping process). The black box with arrows around stands for a process: on the left side are the inputs; on the right side are the outputs; above the arrows control facts are input; and beneath are the help inputs which make the process possible (see figure 3.1)

The set up, the drawing notation and the code system of the IDEF-0 model are standardized.

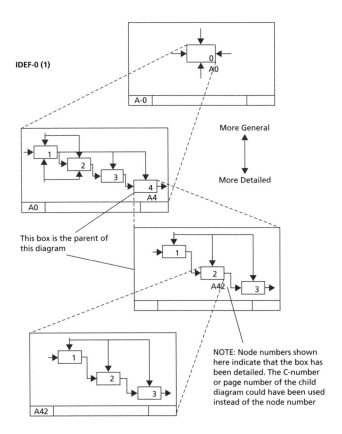

IDEF-0 (1)

A-0

A0

More General

More Detailed

A4

This box is the parent of
this diagram

A42

1

2

3

NOTE: Node numbers shown
here indicate that the box has
been detailed. The C-number
or page number of the child
diagram could have been used
instead of the node number

Figure 3.1 IDEF-0 model

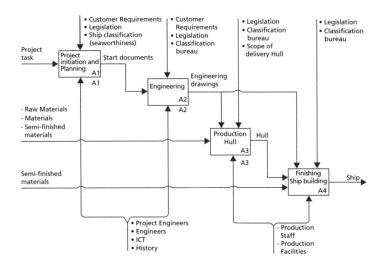

Figure 3.2 Example of IDEF-0 process 'Design of a Ship (2nd level)

3.3.2 Function-product model (Hamburger model)

As already has been mentioned, there can be a big gap between the ideas and wishes of a client and the solution offered by the design team. It is commonplace to have unsatisfied clients at the end of a building process after the building is finally delivered. There are typically conflicts about prices, delivery times, comfort levels in the building, etc. Indeed this picture is not unusual in other fields of the engineering world. In order to overcome this problem it should be possible to create a situation where there is a bridge between the world of the client and the world of thinking of designers. So a good communication between these two parties must be possible.

A model, called the 'functional decomposition model' can be very helpful in closing the aforementioned gap (see figure 3.3). This model is also named the 'Hamburger model' and combines in one model the main function of a product, the functional specifications of a client and the possible solutions. In the model the main function (1st level) is decomposed into at least two sub functions (2nd level), and than again each sub function is decomposed into at least two sub sub functions (3th level), and again these functions are decomposed into at least two sub sub sub functions (4th level) and so on, until the level of separate parts is reached.

At each level a function will become its specifications, the functional specifications, and also there will be at least one solution to fulfil the function and the additional specifications.

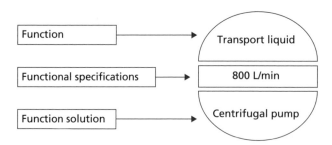

Figure 3.3 Hamburger model

This model is called Hamburger model because in its presentation it can have the shape of a hamburger. The piece of bread above is the function, the meat between represents the functional specifications, and the piece of bread below is the function solution. Mostly there is more than one solution and the challenge is to find the best solution for a given set of functions and the functional specifications.

The main function can often be decomposed into at least two underlying sub functions, which will again have more than one solution. The functional decomposition process can be demonstrated by the pump system, consisting of a main receive tank X, a pump and a delivery tank Y (as shown in figure 3.4).

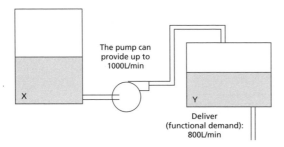

Figure 3.4 Possible solution: a pumping system

How to make a functional decomposition?
We start with formulating the main function, in this case ´to transport water´ (at least 800 litres/min). The chosen solution (an object that can fulfil this function) is a pumping system (see figure 3.4). We know that other solutions are possible (even by hand through a line of people transporting water by means of a bucket). The decomposition starts by investigating and formulating the underlying sub functions, here: ´receive water´, ´move water´, ´deliver water´ with the solutions: ´tank receive system´, ´centrifugal pump (system)´ and ´tank deliver system` (see figure 3.5). For each sub function we investigate and formulate again the underlying *sub sub* functions, for the centrifugal pump (system): ´make a structure´, ´give power´, ´pump water´ and ´control pump system´ with the solutions: frame, electrical motor, centrifugal pump and control system. We can go further with the decomposition so long as we can investigate and formulate at least two underlying functions, and when we come to a point of decomposing we can identify separate parts. In the case of the pump the next step will be: housing, impellor, shaft, bearings, coupling and (mechanical) seal. Now the process of decomposing can stop.

The function – product model is a functional decomposition model which analyzes suitable solutions for a function set (function and functional specifications). Usually a function set has more than one solution, e.g. Function: *Transport Liquid* can be done by a centrifugal pump, a piston pump, a chain of people with buckets, etc.

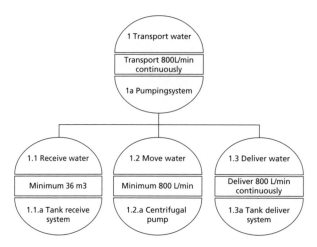

Figure 3.5 Example of a functional decomposition model

Comments on the Hamburger model

The decomposition process of the Hamburger model can be continued for each function at each level as long as this function can then be decomposed into at least two underlying functions. Functions at each level can have more than one possible solution (two, three, etc.).

It is also possible that one solution serves more than one function, e.g. a wall of a building can serve the construction, the sound reduction and the heat transfer reduction (isolation).

Each function (on all levels) and each solution can be given a code number and, with the aid of these code numbers, all types of information can be stored in a structured way in a database. Reuse of knowledge can therefore be enhanced by the storage of this structured information.

Because very often a function has more than one possible solution to fulfil the required specifications, this model can also be applied for value analysis or value engineering studies.

Each function can be allocated or estimated a cost price and its life cycle costs, so that the model can be applied to cost studies (cost price and life cycle costs).

The model can be applied to nearly all of the phases of the IDE process (from ideation through until maintenance). Only during the ERP assemblage phase of manufacturing is it advisable to turn the model 90 degrees to the right (from vertical to horizontal orientation) so that it can be applied as a planning model with a time horizon.

If there is a failure of a function, or in other words a function cannot fulfil its specifications, then we have a situation of functional failure. Functional failures can be used as input for repair activities, or in other words the starting point of the maintenance process. By ranking the functions of each hamburger in the model to a risk scale, it is also possible to evaluate the critical parts in the model.

We will see that this relatively simple model can be applied in nearly every phase of the design life cycle and after this it has a long life during the operation and maintenance phase.

3.3.3 V-model from systems engineering

Systems engineering is very common and is in line with the IDE concepts, in that it is an interdisciplinary approach and a means to enable the realisation of successful systems. It focuses on defining the customer needs and required functionality early in the development cycle, documenting the requirements, then proceeding with design synthesis and system validation, whilst also considering the complete problem: operations, performance, test, manufacturing, cost and schedule, training and support and disposal. Systems engineering integrates all the disciplines and specialist groups into a team effort, forming a structured developmental process that proceeds from concept to production through to operation.

In this world of systems engineering a similar decomposition model to the hamburger model is applied, this model is called the **V-model** (see figure 3.6). The model starts by formulating the main function of a system and the specifications on the left wing of the 'V'. Under the main function are placed the sub functions of the system arrived at by a decomposition process, together with the specifications. This process of decomposition continues to the component level and the part level. The part level is usually reached in four or five steps or levels. All of these functions, sub functions, sub sub functions, component and parts will have their specifications and are placed on the left wing.

On the right wing of the V-model the real solutions appears and the control loop is shown via the (horizontal) arrows. At each level the chosen solution has to show or to prove that the specifications are met. As with the hamburger model, all the functions and solutions can be controlled by a code system, so that information can easily be stored and reused.

Systems engineering principles are applied in surroundings where a lot of companies work together on products such as mobile phones, electronic products, cars, airplanes, etc. These firms work together in a engineering community known as '**simultaneous engineering**', '**collaborative engineering**' and '**concurrent engineering**' under the umbrella of an OEM manufacturer. In the V-model, the contribution of each partner in the design process can be specified.

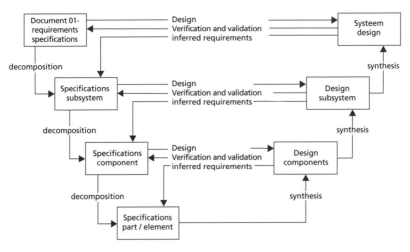

Figure 3.6 The V-model

Hamburger versus V-model

The V-Model has a stronger position in the field of structuring documentation about the system, especially the control side of this documentation with the control loop. However, the Hamburger model is stronger in the way that it presents a function (on all levels) with the possible solutions.

Also with the Hamburger model it is possible to produce an overview of all the possible solutions, or variations in solutions, in one model.

3.4   Code systems

Beneath structure models are **code systems** which are of great importance. Both provide the basic structure for the technical administration system. The product development process produces a lot of data, documents, drawings and decisions, and all of these have to be stored in a structured way for subsequent access and reuse. Unfortunately technical people do not have any real interest in storing data in this structured way and are more attracted to new solutions and new technical challenges.

One of the main reasons we have discovered is that most industrial production organizations have not paid enough attention to the standardization of information systems. Even on one site, departments can use different standards or codes. Unfortunately the importance of a good engineering standard code system is often not recognized. Inherent to the fact that a code system is not seen as primary science is that there is hardly any supporting literature.

As already stated earlier (paragraph 3.2) the selection of a company code (numbering) system is a strategic decision, and is one of the key factors in making ICT systems such as ERP, CAD and CMMS work successfully. The setting up of an engineering code system is not a simple task, a wrong choice results in a lot of problems, whereas a right choice produces successful results. The applied code system has to follow the organization and the type of equipment and is, in practice, always 'tailor made'. There are numerous possible code systems for different organizations and their specific equipment. The decomposition models will be applied for analyzing the equipment.

3.4.1 Hierarchical breakdown model

The **hierarchical breakdown model** (see figure 3.7) gives only the hardware (element) of the function-product model. This is mostly the case in the field of maintenance, where the objects are chosen and known. The need for a decomposition in the form of hierarchical breakdown is recognized as needed.

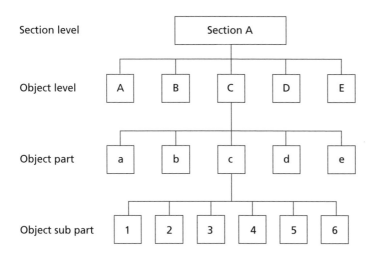

Figure 3.7 The hierarchical breakdown model

The case of the pumping system will be decomposed as follows:

System Decomposition:	Codes decomposition:
A. pumping system	1
Objects of the item 'pumping system´:	
AA. Tank receive system	1.1
AB. Pumping set	1.2
AC. Tank deliver system	1.3
Object parts: (of the pumping set):	
ABa. Frame	1.2.1
ABb. Pump	1.2.2
ABc. Electrical motor	1.2.3
ABd. Control unit	1.2.4
Object sub parts: (of the pump):	
ABb1: Housing	1.2.2.1
ABb2. Impellor and shaft	1.2.2.2
ABb3. Bearing suction side	1.2.2.3
ABb4. Bearing press side	1.2.2.4
ABb5. Seal suction side	1.2.2.5
ABb6. Seal press side	1.2.2.6

Note: The codes can be formulated by letters, by numerals, or by a mix of both.

3.4.2 Code system for the machine shop of an engineering firm

For asset management of the equipment of a machine shop, the data of each machine relating to both operation and maintenance has to be stored. The equipment of machine shops often consists of an assortment of different types of individual machines, each of which have their own function, like: drilling, turning, cutting, etc. If maintenance is necessary on one of the machines, this machine will be taken out of the work planning, and the repair activity can take place. A simple code with place (of the factory), department and unique number is usually enough.

3.4.3 Code system for functions and possible solutions

The main function of the solution Bike is 'To Move One Self'. The main function becomes code 1 and the solution is code 1a, because there are other solutions possible for this function which become the codes 1b, 1c, 1d, etc. We can do the same for the sub functions, e.g. To Structure a Bike', which becomes the code 1.1, with the solution 'Frame of a Bike' becoming code 1.1a.

Generally speaking the function becomes a code, the solutions become the same number and the annexes a, b, c, d, etc are used to distinguish between the different solutions.

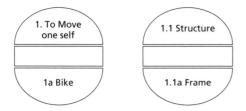

Figure 3.8 The function becomes a code

In scheme form (see figure 3.8) we see the following set up for the code system:

Function:	1	Solution:	1a
Sub functions:	1.1		1.1a
	1.2		1.2a
Sub sub functions:	1.1.1		1.1.1a
	1.1.2		1.1.2a
	1.2.1		1.2.1a
	1.2.2		1.2.2a

Special attention should be paid to how far we go with the decomposition. For product development processes and manufacturing processes we must go to the part level, this is mostly four or five levels down from the main function.

3.4.4 Code systems and organization

In most cases the construction of a code system has to be done carefully and in consultation with all the stakeholders involved in the product development process.

All type of data, notes, drawings, documents, research reports, etc. can be stored with the codes and so retrieved immediately. Therefore, access to all knowledge can now be achieved very quickly and when this knowledge is stored without mistakes it is perfect.

A consequence of an engineering organization working in this way is that it needs to set up a function or department to monitor the structure models and code systems, and allocate the codes on new developments. Also the standardization of the names of solutions (with of course equivalents) and parts has be strictly controlled.

When it is well organized, the content of the databases is very easy to access and the reuse of the existing knowledge is not only facilitated, but also results in lower costs as a result of quicker and failure-free working.

Case

In a design department of a machine factory four design engineers were working on the same 2D CAD system. One of the four left the company and after some time questions about details of his drawings were raised with this department. However the remaining colleagues discovered that it was impossible for them to understand the design details of their former colleague! And it was impossible, therefore, to answer these questions directly.

They also discovered that each of them used the CAD system in their own way. So any communication between the respective CAD applications of these three designers was impossible. What was happening here? The designers had used the CAD system as a purely personal drawing tool with no interaction between the applications. So, one of the great advantages of CAD systems, the ability to electronically interchange drawings and information, was not possible. There were no standards and code systems in use, they did not even realize that standards and codes for information systems are essential conditions for successful applications.

Code Systems and CAD-CAM, PDM, PLM ERP, CMMS

The well known business information systems like CAD, CAM, PDM, PLM, ERP and CMMS are empty boxes which have to be filled with the processes, structures and code models of individual companies. A specific department or function is needed to control the models and the codes.

3.4.5 Structure models and code systems for product development teams

As already stated earlier, in order to monitor the structure models and the code systems it is necessary to have a person or department who is responsible for controlling the decompositions and the code systems. Only one department can authorize the setting up of these models and codes. This also means that as soon as possible when a product enters the phase of functional decomposition within the product development team, one person has the task to obtaining authorization from the department responsible for the standardization of codes. When the set up is authorized, all of the documents associated with the development of this product are decomposed and coded by this system. Also, in the engineering database the information is stored on basis of these models and codes. The advantages are tremendous because everyone can have access to the common engineering data, with the direct result that it takes less time (up to 80% is found) and there are less failures (zero failures are possible).

3.5 Hamburger model as Product Configuration Model (PCM)

When the development of a product is finalized and all delivery modes are known, then it is possible to compose a Product Configuration Model on basis of the Hamburger model. The problem is that all relations between functions and the specifications have to be 'tailor made' by relation matrixes. This has to be done 'by hand'. When the Product Configuration Model is to be placed in an engineering database then a *structurist* with a technical background and knowledge of programming an ICT model must make the first attempt at developing the Product Configuration Model. This requires a lot of special programming on the 'tailor made' Hamburger model in order to translate the content of the engineering world to the ICT world.

This structurist, an extremely valuable person for each company, must possess 70% engineering experience and 30% ICT knowledge, which is a very rare competence! Every design department should have such a structurist to develop the structure in their ICT systems. Such a person is necessary to ensure effective communications between the engineers and the ICT programmers and make sure that the thoughts and wishes of the engineers are understood.

Example:

A function has as a solution a production machine with different capacities A, B and C (a machine family), which again consists of two sub functions and their solutions. Depending on the capacities of A, B and C, the solutions of the sub functions will have different sizes of equipment, which have been chosen in relation to the capacity of the main function. In the relation matrix we can store these fixed relations.

When Capacity A is chosen then transport capacity 'a' is applicable in all solutions. When Capacity B is chosen then transport capacities 'a' and 'b' are applicable in all solutions.
When Capacity C is chosen then transport capacities 'b' and 'c' are applicable in all solutions.

The relations are shown in the matrix form in figure 3.9.

Function /Solution	1a	1.1a	1.2a
1: cap. A	a		
cap. B	a and b		
cap. C	b and c		
1.1: cap. A		a	
cap. B		a and b	
cap. C		b and c	
1.2: cap. A			a
cap. B			a and b
cap. C			b and c

Figure 3.9   Relation matrix

3.6   Management Paradox (2)

Nevertheless, despite the efforts that companies put in when setting up and implementing ICT systems, we very often notice the non-active participation by top management in generating the right process and structure models and code systems for their own organization. These are the essentials for ensuring that ICT systems work successfully.

Summary

In this chapter we introduced some structure models and code systems to make it possible to develop a structured engineering database. Also a product configurator can be assembled on basis of these models. This product configurator is a helpful tool for the commercial departments, such as sales and sales engineering, for putting together commercial proposals and for analyzing client wishes when developing a commercial project proposal.

Each person working in an engineering environment has to realize that when working as part of a team, such a structure is essential in order to achieve the required figures for quality and speed through the application of concurrent engineering principles.

Each team must have such a structurist who can handle the models and codes in line with the company's standard practices.

Exercises

1. Function: *'to make some coffee'*
This function has the following possibilities:

*   Coffee machine;
*   Making coffee in the Arabic way (Turkey, Greek, etc.);
*   Expresso machine;
*   Senseo machine;
*   Dried coffee powder and hot water (Nescafe);
*   Making coffee the Dutch way (boiling water and filtering through coffee powder);
*   Etc.

Making coffee is a process and the solutions depend upon the taste of the consumer.

a. Develop a Hamburger model of the function: *'make some coffee'* with the following specifications:

Function:	To make coffee for 12 cups
Functional specifications:	Light to strong, max. 12 cups, can hold warm at 80 °C for two hours
Solution:	Home coffee machine

b. Make an IDEF-0 model of this function:

On the main level Ao the specifications are as follows:
Input: amount of water (for x cups), amount of coffee (for x cups, light, middle or strong), filter bag.
Output: x cups in container at temperature y, filter bag with processed coffee powder.
Control: temperature hot water, water level water container.
Help: electrical power to drive the processes.

Literature

- IDEF-0, see www.itl.nist.gov/fipspubs/idef02.doc: Draft Federal Information (Processing Standards Publication 183) 1993 December 21, the Standard for INTEGRATION DEFINITION FOR FUNCTION MODELING (IDEF-0).
- Systems Engineering, see www.incose.org
- Zaal, T.M.E., Lecture Notes of IDE (Code Systems), 2004-2007, Hogeschool Utrecht.
- Clausing, D., *Total Quality Development, A Step by Step Guide to Word Class Concurrent Engineering*, ASME Press, 1998, ISBN 0791800695.

Chapter 4
How to Map the Client's Demands and Wishes

4.1 Introduction

Mapping the correct wishes of the client is one of the most difficult phases in the whole design process. A lot of marketing tools and road mapping tools are available to scope these wishes. On the other hand new products are still created out of the blue by a creative wave without applying any tool. In this book the design process will start at the moment that a first form of a product definition is ready, and the client demands and wishes about a product are more or less clear. The so called pre-initiative phase, including lots of creative possibilities, is not the main scope of this book. On the other hand the client wishes and expectations compared to the product that is actually designed shows how misunderstandings do occur. The well known pictures included in figure 4.1 illustrate this process very well.

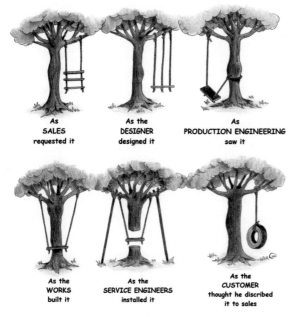

As
SALES
requested it

As the
DESIGNER
designed it

As
PRODUCTION ENGINEERING
saw it

As the
WORKS
built it

As the
SERVICE ENGINEERS
installed it

As the
CUSTOMER
thought he discribed
it to sales

Figure 4.1 What went wrong?

When we look to the **IDE company model** (see figure 4.2) the development work takes place on the development plan. Mostly ideas from new products come from the strategy plan and/or from the innovation agenda. In order to execute the product development work, integrated design and engineering teams have to be put together. We all have to realize how carefully this composition work has to be done.

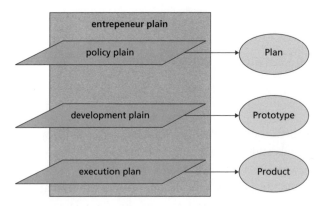

Figure 4.2 The IDE company model

A good product definition also includes an idea about the market size of the new product, about the selling price and a overview about client demands and wishes. There is a difference between demands and wishes. Demands *must* be included in the product and wishes *may* be available in the product. For clients the real difference between products is the amount of wishes that are included for a certain price.

From the beginning of the design process a functional decomposition model will be used to list the function and the sub functions of the product, and one or more possible solutions will be generated for each defined function in the product. The functional decomposition of a product will also generate a structure model for CAD and 3D systems. With a technique such as Quality Function Deployment (QFD) these functions and solutions can be analyzed on their applicability and feasibility. The outcome of this chapter will be a product definition of a product, on the basis of functional specifications, in the form of a functional decomposition in which the functions and function solutions are tested on feasibility. In addition, a first idea will be given about the frame work of a product definition team.

4.2 Developing or Creating Business with Clients

Clients can bring in new ideas about existing products, new products and new fields of operations. When we look at the business model (figure 4.2) we see three main plains in which a business takes place. Brand new fields of operation belong to the policy plain, new products (in line of the existing ones) to the policy and development plains, and the new ideas about existing products (improvements on outcomes, or new functions) belong to the execution plain. Creating business with clients is often a fruitful and profitable way for developing client satisfaction and a long term relationship. A lot of firms utilise client forums as input for continuous product development and improvement.

4.2.1 Example demands and wishes

A good example of the difference between demands and wishes is the process of buying electronic goods by one self. Going to one or more shops, you primarily have in mind a price and an idea about qualities or options. Once you arrive in the shop, the process of comparing starts. Product A has certain features, product B has more features at the same price, whilst in another shop product C has the same features as product A but at a lower price. At the end of the day the choice is between a product with a lower price than you had in mind and the right features, or another at the price you had in mind but with more features. In both cases you will have a good feeling because the price quality ratio gives more than you expected. In practice one is using only 20% of all possible options, so this whole range of options is only used for selling us a product!

For a product designer it has to be very clear what the client demands are and what the client wishes to pay for the product, and which wishes are distinctive to other competitors.

4.2.2 Client wishes and functions

Clients think in terms of functions and designers in terms of solutions. Very often a designer already has a solution in mind when they hear the client wishes. The question is, does this solution really match all the client wishes? Very often it appears that this is not the case, so the next question will be how to make this gap between client ideas about their wishes and the solution(s) in the mind of the designer as close as possible.

At first the client will come in the picture. The client usually thinks functionally about a product, in terms of what the product can 'do'. The consequence of this way of thinking are that the design team has to list all the 'do's' of the functions and also has to weigh the importance of each 'do' for a client. It is very important for the definitive product definition document to understand the ´do´s´ for which a client will want to pay and the ´do´s´ which they are expecting to be included within the price. Then the designers, having generated ideas about the possible solutions, can fulfil the client wishes.

Functions always consist of a verb, which express what has to be done: to mix, to drive. The solution always is a noun, which executes what has to be done; mixer, car, etc. The functional specifications give the boundaries within which the solution has to operate. For example, the capacity of the mixer in kilograms, the maximum speed of the car in kilometers/hour, etc.

4.2.3 Client thinking in the functional manner

To explain the way of client thinking in the functional manner, three examples are shown below:

1 **Function:** *´to move from a to b´*
At first, the example: *´to move from a to b´* as a function will be used. Looking at this function definition the following solutions are possible: boat, aeroplane, car, train, bike, motor cycle, even walking, etc.
A function together with some further specifications always gives limitations in terms of the possible solutions. See the next possibilities:
- When the specification speed is added, e.g. > 5 km/hr, each solution is still possible, whilst at a speed of > 500 km/hr only an aeroplane is possible.
- When to move ´dry´ is added then the solutions bike, motor cycle and walking become impossible.
- When a maximum price: €750 is added, then only the bike and walking are possible.
- When money plays a main role the solution of walking is always the cheapest one.

Hamburger model for a bike
Function: to move (one self)
Functional specifications: at a speed higher than 15 km/hr
Possible solution: bike

The solution 'bike' can be recognized or found in at least five sub functions on the second level: to built parts together, to drive, to manage the direction, to brake, to have safety, etc.

The possible solutions are: frame, crank set, handle bar, brake system, light and reflector (see figure 4.3).

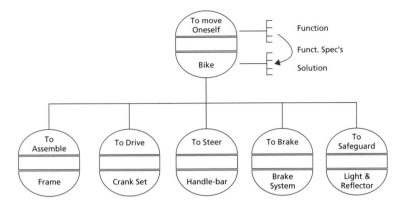

Figure 4.3 Hamburger model for a bike

Again each solution can be decomposed further in new sub functions. The decomposition will dissolve in single parts mostly on level four or even five.

2 Function: ´to make light´
Another example of a function is ´to make light´.

Looking at this function definition, the following solutions are possible:

- A candle;
- An electric lamp (energy by battery);
- An oil lamp;
- A gas lamp;
- Etc.

When the specification easy and safe to move is added: an electrical lamp set on battery power is chosen.

Functional Specifications are: minimal X Lux and minimal time to burn is Y hours continuously.

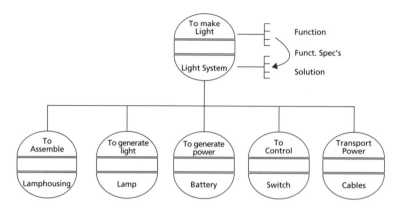

Figure 4.4 Hamburger model light system

3 Function: 'live in a house'

The function 'live in a house' has a lot of sub functions for characterizing the way of living in a house:

- to live   living room
- to cook   kitchen
- to dine   dining room
- to wash   toilet or bath room
- to sleep   bed room
- to store   warehouse
- to play games   games room
- to crop flowers   garden
- to swim   swimming pool
- etc.

Possible types of houses are: villa, detached house, apartment, caravan, bungalow, terraced house, semi-detached house, etc.

The selected house is dependent upon the amount of money, the desired sub functions and the type of house (and also the location).

Hamburger model of a house
Function: to stay or live in a sheltered surrounding.
Functional specifications:

- number of people (adults and children, guests);
- to be able to live, to eat, to cook, to sleep, to wash and clean, to store, etc.

Solution: house. Other possible solutions: shelter, caravan, cave.

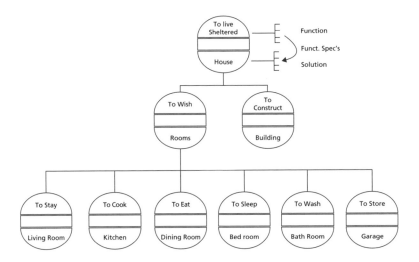

Figure 4.5 Hamburger model of a house (first, second, and third level)

In the building environment we see the origin of a *double hamburger model*. On the left side of the model is the world of the client with their wishes about the rooms and what to do in these rooms, whilst on the right side is the world of the builders with their often purely technical decomposition of the building. With the help of the modern 3D CAD systems and internet technology it is possible to create a virtual world so that a client can walk through their new house and discuss, via this virtual world, what their wishes and concerns are directly with the builders (see figure 4.6).

Figure 4.6 Visit and adjust in the virtual world the house you maybe interested in to buy (www.woonwebsite.nl)

4.2.4 Mapping client wishes

Before we can start the process of mapping client wishes, we have to realize that frequently there are perhaps two or more options they are considering, or the product they have in mind is totally new, so the client does not exactly know what the behavior of the product will be, or if the product already exists and a new version will be developed. An example of a new product is the IPod (MP3, 4 players) and an example of an existing product is the car, with a basic form that is already over hundred years old!

Why this distinction between new and old products? For new products, clients do not exactly know what the real benefits and possibilities of this new product are. So the team that has to address the wishes of the client has a more difficult task than a team that is simply involved in improving or redesigning an existing product. Especially when it comes to evaluating the importance of functions, or missing potential functions. An example of missing or under estimating potential functions is the development of video recorders at the Philips Company. They were on the market with recorders about five years before the Japanese competitors came! All marketing studies indicated a wish to show movies in a easy way through pre played films on cassettes. The development and marketing teams ignored this wish completely and concentrated on educational applications, and on the substitution of the 8 mm small film for making home and holiday family movies. The outcome is well known in that the Japanese firms came with pre played film cassettes and won the battle by distributing these films through video libraries and rental companies, in particular films with adult themes! The Philips Company lost this battle completely.

Sometimes a new product appears out of the blue without any market research and is still very successful. An example is the worldwide success of the ice lolly 'Magnum' (from Unilever, Ola, etc.), the origin of which came from a surplus on production capacity. Without any real marketing campaign this 'by-product to solve a production problem' was launched and the product nearly sells itself.
The car business also has its minor products. Very often the styling is a point, as can be image or sentimental wishes rather than the quality of the car itself.

A particular point of attention is the question: 'does a client know all their wishes and can they formulate these wishes, especially for completely unknown products or services'? Well known in Europe is the 'UMTS disaster' of some years ago. Clients did not see the benefits of, or did want to pay for high speed data communications via mobile phone. Only the simple low cost service SMS was, and is still, a very successfully application.

How can we organize the mapping of client wishes or identify client needs and benefits?

Several methods are known and briefly we will mention them here:

- *brilliant idea*: having a brilliant idea out of the blue for a brand new product is possible and will always occurs at unexpected moments;
- *brainstorming*: a well known methodology to list the possible client ideas (wishes, needs, benefits);
- *interview customers*: the method of interviewing is often the starting point for identifying the client wishes. Client wishes are often not directly clear and are also often functional formulated (by verbs);
- *crowdsourcing*: brainstorming via the internet and user generated content, a new form of brainstorming of invisible clients on internet.

Case: User generated design

'Designed by you, produced by us' is the message of RYZ. RYZ has introduced a new way of thinking about the shoe design business that includes getting input - and ultimately forming a partnership - with their customers. RYZ invites anyone to design the perfect pair of shoes, and they produce the most popular designs, sell them around the world, and share the profits with the designer. How does it work? You can download a template and design your shoe using it. Upload your design to your personal workspace on RYZ's website. People can vote on the design. The winner gets $500 and $1 royalty on every pair of shoes RYZ sells on the internet.

Figure 4.7 Design your own shoes (http://www.ryzwear.com)

4.3   Quality Function Deployment (QFD)

Quality Function Deployment (QFD) is a methodology for structured product planning and development that enables the clear specification of the client's wishes (wants and needs) or the evaluation of each proposed product systematically in relation to its impact in meeting those wishes. It originates from Japan and is also widely used in the USA (D. Clausing and L. Cohen are well known promoters of this method) and in Europe nowadays.

The QFD process involves constructing one or more matrices. The first of these matrices is called the **House of Quality**, and it displays the client wishes (wants and needs in the so called 'Voice of the Customer'). In figure 4.8 the House of Quality (HoQ) with its main rooms 1 to 8 is presented.

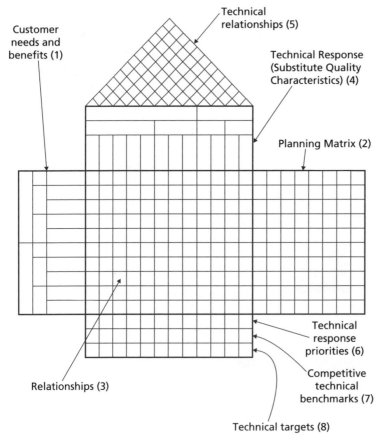

Figure 4.8 House of Quality (HoQ)

When mapping the client wishes, the key questions are always how to organize the mapping of these wishes, how to structure the wishes and how to prioritize these. In chamber 1 of the House of Quality (HoQ), methodologies are developed to deal with and handle a major part of these questions.

In the QFD methodology the list of client wishes has to be processed in a several steps by means of a structured form, the so called **Voice of the Customer**, in chamber 1 of the HoQ.

These steps mentioned are:

- customer needs: in this step the client wishes are collected, sorted into major categories (levels), prioritized and structured in a tree form (see Hamburger model);
- functions: the client wishes are also transformed in functions;
- reliability: the functions become a reliability figure;
- target value: a measurable figure;
- Substitute Quality Characteristics (SQC): a weighing factor for client satisfaction(s).

The Voice of the Customer is a document with an overview of all possible clients wishes, often very softly formulated and always based on qualitative data. This data is structured in a tree form and always given a form of priority.

The next step will be weighting and prioritizing these clients wishes in a range of categories, in the so called **Planning Matrix**, Chamber 2 of the HoQ (see figure 4.8).

These categories are:

- Importance to customer: this is the place to record how important each wish (need and benefit) is to the customer. Ranked from 'not at all' to 'of highest importance'.
- Customer satisfaction performance: this is the customer's perception of how well the current product is meeting their needs. Ranked from 'very poorly to very well' and 'does not apply'.
- Competitive satisfaction performance: in order to be competitive, the development team must understand the competition. Many development teams do not study their competition and often operate in the dark with regard to their competition.
- Goal: the development team decides what level of customer performance they want to aim at in meeting each customer need. Ranking from '1 to 5'.
- Improvement ratio: the goal rating combined with the current satisfaction rating is used to set this. Improvement Ratio = Goal / Customer Satisfaction Performance.
- Sales point: this contains information characterizing the ability to sell the product, the most common values are; no (1,0), medium (1,2) and strong (1,5) sales point.

- Raw weight: this factor contains a computed value from data and decisions made in the above categories of the planning matrix. It models the overall importance of each customer wish. Raw Weight = (Importance to Customer) × (Improvement Ratio) × (Sales Point) or Raw Weight = (Importance to Customer) × (Goal / Customer Satisfaction Performance) × (Sales Point).
- Normalized raw weight: contains the Raw Weight values as a percentage of the sum of the Raw Weights, so Normalized Raw Weight = Raw Weight / Raw Weight Total.
- Cumulative normalized raw weight: by making a Cumulative Normalized Raw Weight column. It is useful to rank the different Normalized Raw Weight from the highest value to the lowest.

The outcome of the Planning Matrix (see figure 4.9) is a matrix with estimated and calculated figures, which give a ranking of the importance to the client of all the wishes of the Chamber 1 customer needs (Voice of the Customer). For the development team this outcome is an important part of a product definition document for a new product.

Note: The figures in the Planning Matrix mostly come from Marketing departments, but the choice of each factor is often done on basis of feeling(s) and this fact can give the outcome a possibility of mis-ranking. The Philips story of the video recorder shows this case. They knew of the wish to show pre-recorded films on the TV set via the recorder. When the Japanese firms came on the market roughly five years later with recorders and the pre-recorded films (especially the more erotic ones!) the battle in the market for a world standard was lost in no time by the lack of this form of feature!

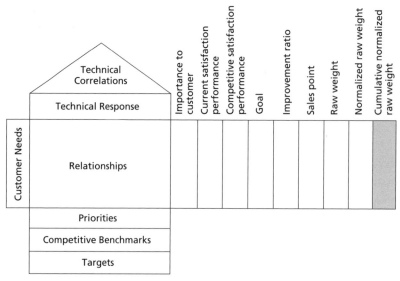

Figure 4.9 HoQ with Planning Matrix

4.3.1 How to fill a Planning Matrix

To illustrate how to fill a Planning Matrix we take a part of an example from the book of L. Cohen. This example deals with client wishes (or wants and needs in the House of Quality) in relation to a new to develop text program for a computer surroundings with the next points or wishes to evaluate:

The program is a pleasure to use:

- Commands are easy to know and use:
    - Intuitive controls;
    - Controls under my fingertips;
    - Can customize to suit my working style;
    - Easy to get the information I need.
- Program is quick and responsive:
    - Can adjust the cursor to move as quickly as I should like;
    - Enables me to find things in the document quickly.
- Easy front management:
    - Offer a lot of size, font and design options;
    - Able to see what the fonts look like as I am choosing them.

For filling in the Planning Matrix we will apply the factors and weight ranges mentioned in table 4.1.

Factor	Minimum Value	Maximum Value
Importance to Customer (Absolute)	1	5
Importance to Customer (Weighted)	1	100
Customer Satisfaction Performance	1	5
Competitive Satisfaction Performance	1	5
Goal	1	5
Improvement Ratio = Goal / Our Current Rating	0.2	5
Sales Point	1	1.5
Raw Weight (with Absolute Importance)	0.2	37.5
Raw Weight (with Relative Importance)	0.2	750
Normalized Raw Weight = Raw Weight/ Raw Weight Total	0.01	0.99

Table 4.1 Weight ranges in the Planning Matrix

With the help of the factors of weight ranges we can fill the Planning Matrix for this case (see table 4.2).

Note: The team members mostly choose the factors to fill in so a possibility of subjectivity can occur. It is advisable to let fill in a part of these factors in the planning matrix by client panels.

	Weighted Importance to Customer	Customer Satisfaction Performance	Competitive Satisfaction Performance	Goal	Improvement Ratio	Sales Point	Raw Weight	Normalized Raw Weight	Cumulative Normalized Haw Weight
Can customize to suit my working style	81	4.6	3.8	4.6	1.00	1.5	559	0.19	0.19
Easy to get the information I need	80	4.7	4.6	4.7	1.00	1.2	451	0.16	0.35
Controls under my fingertips	83	3.1	4.4	4.4	1.42	1.2	438	0.15	0.50
Intuitive controls	84	2.9	2.8	3.3	1.14	1.5	416	0.14	0.65
Enables me to find things in the document quickly	48	3.1	4.4	4.5	1.45	1.5	324	0.11	0.76
Offers lots of size, font, and design options	45	4.6	3.8	4.6	1.00	1.5	311	0.11	0.87
Able to see what the fonts looks like as I'm choosing them	42	4.7	4.6	4.7	1.00	1.2	237	0.08	0.95
Can adjust the cursor to move as quickly as I'd like	49	2.9	2.8	2.9	1.00	1.0	142	0.05	1.00
Totals							2878	1.00	

Table 4.2 Completed Planning matrix

We see that 'Can customize to suit my working style' has the highest score 0.19 (normalized raw weight) and that 'Can adjust quickly the cursor to move as quickly as I'd like' only becomes 0.05 points.

4.4   Customer Satisfaction and the Kano Model

Noriaki Kano, from Japan, has developed a model of customer satisfaction (see figure 4.10). The essence of the so called **Kano model** is the relation of customer satisfaction to product characteristics. These product characteristics are divided in three distinct categories, which are:

- *Dissatisfiers*, also known as 'must be', 'basic', or 'expected' characteristics;
- *Satisfiers*, also known as 'one-dimensional: or "straight-line' characteristics;
- *Delighters*, also known as 'attractive' or 'exciting' characteristics.

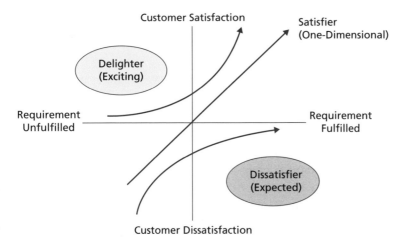

Figure 4.10 Kano model of customer satisfaction

Dissatisfiers are characteristics that clients expect to be present. Satisfiers are things that clients want in their products (to buy). The more satisfiers that are provided, the happier the client will be. Delighters are product attributes or features that are pleasant surprises to clients when they first encounter them. Delighters are very important because clients will not be dissatisfied if they are not present because they are unaware of what they are missing! On the other hand delighters are called 'exciting quality' or 'unexpected quality', and typically have 'Wow' characteristics.

During the life time of a product a delighter can move to a dissatisfier. An example is the air condition system for cooling and heating in European cars. Ten years ago this was a delighter, whereas now it is normal and common for each car, and has become a real dissatisfier that is expected on each car. Satisfiers are characteristics that give a better feeling, for example a price lower than expected or a higher capacity higher than expected, etc.

4.5   QFD and the Other Six Chambers

We have discussed the content of the chambers 1, and 2 of the House of Quality (HoQ) within the QFD methodology, in which the Voice of the Customer is heard, and out which results an overview of weighted and prioritized client wishes.

Now we will translate these client wishes into technical solutions or possible products. We will also apply the methodology of Kano with his distinction between dissatisfiers, satisfiers and delighters during the product definition process.

When we look in some detail at chambers 4, 3 and 5 and later at chambers 6, 7 and 8, we can complete the House of Quality (see figure 4.8). We follow again the methodology of QFD, as described by Lou Cohen in his book, in a condensed form.

In the same way as the Voice of the Customer (VoC) had a qualitative component (see left, chamber 1 customer's needs and benefits) and a quantitative component (see right, chamber 2 Planning Matrix), the translation of the VoC to the **Voice of the Developer** has a qualitative and a quantitative component. The translation is also called '**substitute quality characteristics**' and has the qualitative part in the *upper* chamber 4 (technical response) and the quantitative part in the *under* chambers 6 (priorities), 7 (benchmarks) and 8 (targets). Finally, chamber 3 is the relationship chamber between the different chambers. We start with chamber 4.

4.5.1   Chamber 4: technical response

In upper chamber 4 of the HoQ the development team's technical response to the customer's needs is described. It is the translation of the **Client's Quality Characteristics** into technical terms, the **Substitute Quality Characteristics (SQC)**; effectively from the customer's language into technical language. This **technical response** describes a product from any one of a variety of points of view, also called the product or the design requirements. In practice this means a translations of the *whats* of the customer into the *hows* of the designer, (see figure 4.8). In a matrix form the *whats* will stand on the left side, and the *hows* on the topside. Each *what* (or customer attribute) can be translated in one or more (technical) performance measurements (see figure 4.11)

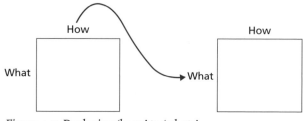

Figure 4.11 Deploying 'hows' to 'whats'

These measurements can be measured properly and can be controlled by the development team. They should be characterized in a few ways.

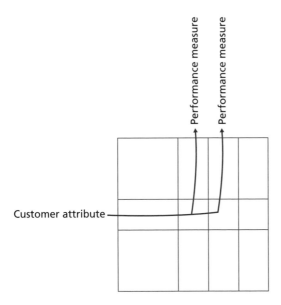

Figure 4.12 Customer Attribute Deployed to Performance Measurements

First the units of measurement will be defined. E.g. voltage in volts, time in seconds, etc.

Second the *direction of goodness* will be defined on basis of three directions of goodness:

- The more the better, e.g.: reliability (by mean time to failure);
- The less the better, e.g.: quality of service (measured by number of defects);
- Target is best, e.g.: constancy of temperature in a food freezer container.

The next step in the process is to define the measurements by describing how each measurement will be performed. Also all assumptions and comments about each measurement should be documented, e.g. the RAM specifications of each function. For designers this step in the process will be often be ignored because they think that such measurements are so self- evident that they do not need an explicit description. Such an omission can lead to great troubles later in the development process e.g. during the testing of the specific SQC's.

Other way: product functions and function trees (Hamburger model)
There is a completely different way to define the Substitute Quality Characteristics. It is also possible to place *product functions* along the top of the HoQ. However, many products have large numbers of capabilities or functions, so it is not realistic to place them all at the top of a HoQ. A decomposition model, like the Hamburger model, or in general a **function tree**, makes it possible to rank these functions to several levels and place the SQC's on all these levels. So the process for chamber 4 can be done by the translation from the VoC to performance measures, to functions (and function tree), to product design. Each pair of topics will represent the left and the top of a new matrix (three matrixes).

After formulating the main function of a product, the primary subsystems of this product will be identified, and so on for the secondary subsystems. Each of these subsystems (and components) will be described at several levels of additional detail.

4.5.2 Chamber 3: relationships, impacts and priorities

Again chamber 3 is used to show the relations between the customer needs and the formulated Substitute Quality Characteristics in such a way that the priorities of all these SQC's can be developed. Chamber 6 is the priorities chamber, so the impacts of the relations can be prioritized. Again matrixes are used to study these relationships.

The first matrix gives the relationship of the customer needs to the technical responses (SQC's). The impact of the possible relationships ranges from *not linked*, via *possibly* and *moderately* to *strongly* linked. Symbols and numeral values symbolize the impact of the relations (see figures 4.13 and 4.14).

Symbol	Meaning	Most Common Numerical Value	Other Values
	Not linked	0	
△	Possibly linked	1	
○	Moderately linked	3	
◉	Strongly linked	9	10, 7, 5

Figure 4.13 Impact symbols

The second matrix calculates the relative weight of importance of each SQC to the total or overall customer satisfaction. The applied numerals for calculating this importance vary from zero (not linked) via 1 (possibly) and 3 (moderately) to 9 (strongly linked). The result is shown in figure 4.15.

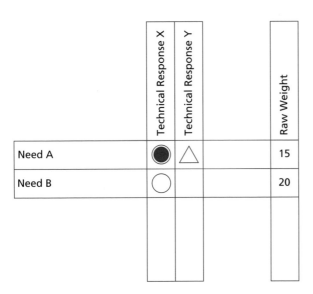

Figure 4.14 Amount to impact

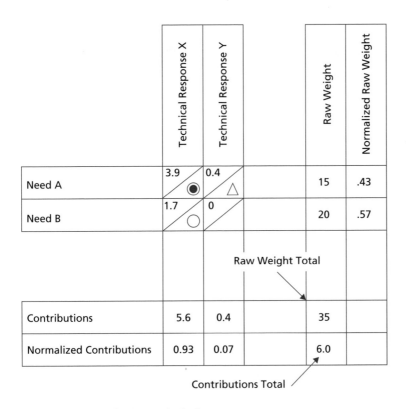

Figure 4.15 Contributions calculations

The outcome of chamber 3 is shown in figure 4.16.

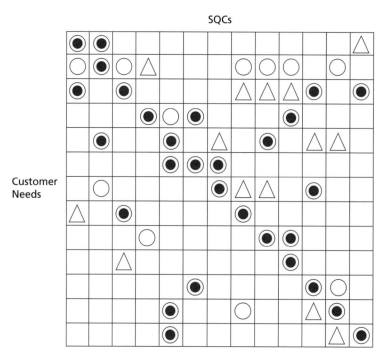

Figure 4.16 The pattern of impacts

4.5.3 Chamber 5: technical relationships

Chamber 5 incorporating the **technical relationships** is the roof of the HoQ and contains the correlations between the SQC's. Some SQC's even have a negative correlation between each other, whilst others have a none or a positive correlation. E.g. for an automobile the air conditioner gives better climate (SQC_a, the more is better) but also a higher weight (SQC_b, the less is better), so a negative impact.

The correlations section consists of that half of a matrix that lies above the matrix's diagonal. The SQC's are arrayed along the top and the side (see figure 4.17). Then the matrix is rotated 45 degrees and so will form the roof of the HoQ. For those SQC's classified as satisfiers, the team can expect that the better they perform on the SQC, the greater the customer satisfaction will be.

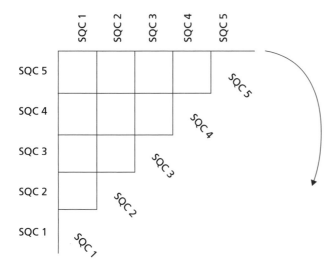

Figure 4.17

The targets (both numeric and non-numeric) are expressed in the categories: development team's target, best in world, target (chosen). Sometimes best in world and other times the development team's value is chosen as target, this choice can be lower, equal or even higher than world class (see figure 4.18).

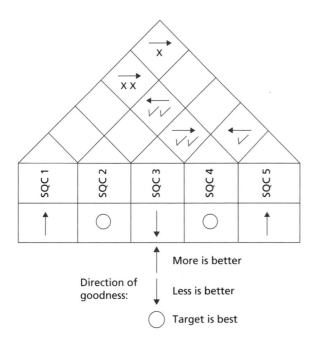

Direction of goodness:

↑ More is better

↓ Less is better

◯ Target is best

Figure 4.18 Roof of the HoQ

The weighting of the correlations between the SQC's five degrees of technical impact can range from strongly negative, moderately negative, blank, moderately positive through to strongly positive impact (see figure 4.19). In this figure the direction of the impact is also given.

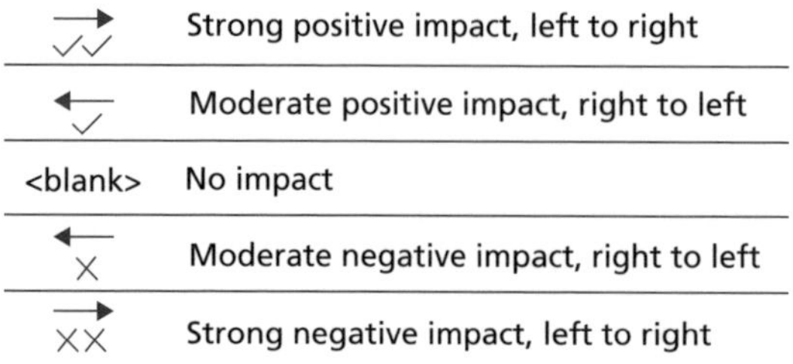

$\overrightarrow{\lor\lor}$	Strong positive impact, left to right
$\overleftarrow{\lor}$	Moderate positive impact, right to left
<blank>	No impact
$\overleftarrow{\times}$	Moderate negative impact, right to left
$\overrightarrow{\times\times}$	Strong negative impact, left to right

Figure 4.19 Degrees of technical impact with directions of impact

With the help of the technical correlations it is also possible to indicate which teams or members of the design team must communicate with each other during the development process. To make these communication correlations more explicit a **responsibility matrix** can be constructed, again guided by **responsibility symbols** (see figure 4.20). An example of a responsibility matrix is given in figure 4.21 so the team members can see how they are correlated to the different SQC's.

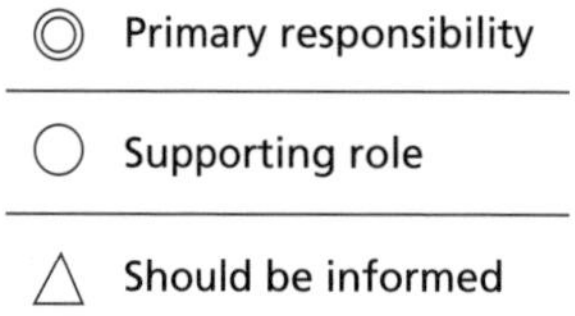

◎	Primary responsibility
○	Supporting role
△	Should be informed

Figure 4.20 Responsibility symbols

	Organization A	Organization B	Organization C	Organization D	Organization E	Organization F	Organization G
SQC 1	◎		△		△		
SQC 2			◎			○	
SQC 3		◎		○	△		○
SQC 4			△		◎		
SQC 5			◎		△		

Figure 4.21 Responsibility matrix

4.5.4 Chamber 6: Priorities

In chamber 6 the key SQC's are prioritized following the investigations in chambers 4 and 5. The SQC with the highest customer satisfaction becomes the highest priority, etc. (see outcome in figure 4.15).

4.5.5 Chamber 7: Benchmarks

In this chamber the competitive technical benchmarking has it place. Every design team must look at some point at the competition in the outside world. Two types of benchmarking are investigated for the key SQC's: benchmarks for performance measures; and benchmarks for functionality.

For performances measures the benchmarking process becomes one of measuring the competition's performance and one's own performance in terms of these measures. For functionality the product or service functions are explicitly defined, and consequently the benchmarking process is more subjective because the competition's design of a SQC is mostly different to that of the design team.

Many methodologies are available to assist in the benchmarking process and, depending on the experiences of the design team, a methodology can be applied.

4.5.6 Chamber 8: Targets

In this chamber the process of setting targets for the key SQC's is undertaken (see the outcome of chamber 6 'priorities' and chamber 7 'the benchmarked competition'). There are two types of targets: numerical targets and non-numerical targets.

The QFD process itself does not provide a cookbook approach for setting targets for SQC's. The outcomes will depend to a degree on the business know-how and technical expertise of the development team. The target setting stage is also a good place to deal with the Kano classifications, so it is very useful to classify these key SQC's according to their category in the Kano model.

For those SQC's classified as potential delighters, the team must decide how aggressive it will be in target setting. Remember that customers will not notice the absence of a delighter!
For those SQC's classified as dissatisfiers, the team cannot afford not to be aggressive. Customers expect perfection in these areas.

Development teams must be aware of the feelings of clients about each product in relation to what are the dissatisfiers (the client expects these as normal), what are the satisfiers (these will give a client good feelings about a product) and what are our surprises, the delighters or the WOW feelings about our product, which will also make the differences between us and our competitors.

By completing chamber 8 it is possible to make the complete QFD document, as part of the product definition document. This document will be applicable to the engineering of the possible solutions of all functions on each functional level.

Note: For a full description of the QFD methodology and the process for its use and implementation see, for example, the book QFD by Lou Cohen, or EQFD by Don Clausing.

Example of a QFD House of Quality

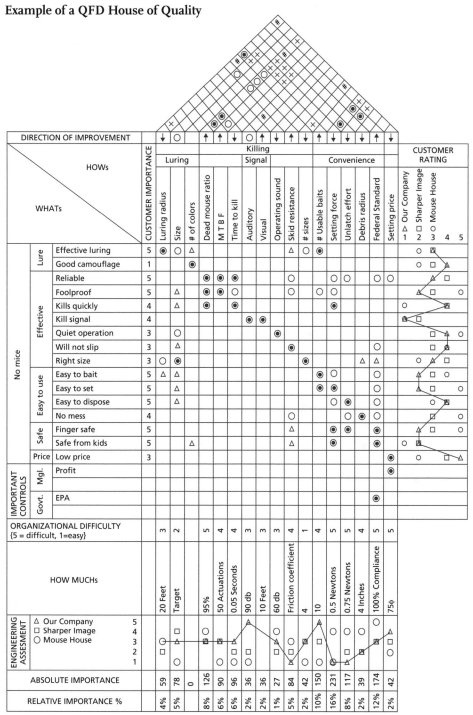

The Mousetrap House of Quality, created using QFD Designer Software, Qualisoft™ Corp.,
4652 Patrick Road, West Bloomfield, MI 48322

Figure 4.22

4.6   Product Definition Document (PDFDoc)

After considering client wishes, it is now time to formulate the outcomes of a **Product Definition Document (PDFDoc)**. Product development teams can apply this document as starting point for the development of the real product. Putting this PDFDoc together is, in practice, the first step in the development process for developing a new product. In order to put the document together a team of marketers and designers will be formed. If possible, some leading clients should be involved in the first part of developing the document.

It is vital to understand what the client really wants, what are the wishes (needs, benefits, etc.). From the Kano model we learned that client always have hidden wishes or qualities for products, the delighters, often unspoken but coming to the surface when one of the competitors offers one or more of these characteristics. The Philips Video Recorder case shows the lack of the delighter of a player already offering finished software in the form of films, and the Unilever Ola Magnum Ice case shows a delighter in the form of an ice cream that is very acceptable for adults.

The team begins with the very important step by organizing the Voice of the Customer of the QFD model with the chambers 'Customer Needs' and 'Planning Matrix'. This provides the input of all client wishes (needs and benefits) from market research, interviews, etc. and these wishes are then placed in a tree structure. Furthermore a weighting and ranking system for assigning the importance of each wish is the outcome of the planning matrix. The Hamburger model will be very useful for generating this tree structure.

The document should also contain a section containing information about competitors, competitive products, and competitive functions for these products.

Further information should also be gathered about the dissatisfiers, satisfiers and possible delighters of the new product.

Finally, the marketing department should give information about markets and prices for the new product.

4.6.1 PDF document

The PDFDoc should contain at least the following parts:

- Business case or market situation for the new product, with estimates on: market potential, competitors, volume of products to be sold, selling process, investments, cost prices, etc.;

- From QFD the Voice of the Customer, the outcomes of chamber 1 'Customer Needs' and chamber 3 'Planning Matrix';
- The outcomes of chamber 8 with all the weighed specifications, the key SQC's;
- Needs formulated in a structured tree form (the Hamburger model);
- Overview of dissatisfiers, satisfiers and possible delighters in relation to the outcomes of client interviews.

Concurrent Engineering Planning (1)

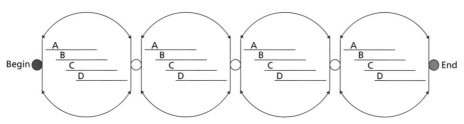

Figure 4.23

After the first phase the PDF Doc is finalized and the next step will start.

Notes:
- Even with all the work associated with formulating the right product definition for a new product, the outcome can be still uncertain. For example, see the role of delighters when launching new products.
- Most development work is done by improving or renewing existing products, and the right balance between dissatisfiers, satisfiers and possible delighters has to be found.
- The question 'who are our competitors?' is very important. Too often development or design teams ignore this question!
- In the PDF Document a lot of choices and decisions are made, which have a great influence on the cost price and the market potential of the new product. Top management have to be involved very actively during this last stage and will the go /no go decision for the next step the complete design and engineering of the new product physically.

Summary

The translations of client needs and wishes into technical solutions for a new product is one of the most difficult engineering problems to solve because beneath pure technical affairs there are also a number of image and material issues that play a role in the satisfaction of clients. The model of Kano shows this problem, in particular his reference to the role of delighters. The models of QFD and Function Product Model (Hamburger model) that we have discussed can be very helpful for the product development teams in clarifying the needs and wishes in relation to possible solutions.

Exercises

1. What might be a delighter for the function 'make coffee'?
2. Make a Hamburger model for a Kart, what is the function of a Kart, what are real functional specifications of a Kart?
3. Make a Planning Matrix (HoQ) of a zero emission (CO_2) power set for the Kart (see figure 4.24).

Figure 4.24 Design of the Zero Emission racing Kart (© Renderhouse bvba, Ronse). A fuel cell converts hydrogen and oxygen into electricity that gives the power to drive. The only emission is pure water, H_2O (Solvay Umicore Zero Emission Racing team, http://www.formulazero.be)

Literature

- Cooper, R.G., *Winning at New Products. Accelerating the Process form Idea to Launch*, Perseus Publishing, 2001, ISBN 0738204633.
- Cohen, L., *Quality Function Deployment. How to Make QFD Work for you*, Addison-Wesley Longman Inc. 1995, ISBN 0201633302.
- Clausing, D., *Total Quality Development, A Step by Step Guide to World Class Concurrent Engineering*, ASME Press, 1998, ISBN 0791800695.
- Pugh, S., *Total Design*. Addison-Wesley, 1991, ISBN 0201416395.
- Ulrich, K.T. et al., *Product Design and Development*. McGraw-Hill, 2004, ISBN 0072471468.
- Zaal, T.M.E., Lecture Notes on IDE (Functional Models), 2007, Hogeschool Utrecht.

Chapter 5
Methodology for Product Design over the Life Cycle

5.1   Introduction

In chapter 4 we discussed the difficult world of mapping client wishes (needs and benefits) in relation to the design of a product. In particular, the case of a totally brand new product which a client has not seen before and does not know anything about. It gives a design team a lot of uncertainties about the real importance of each wish on the list and how well the client will receive the final result.

From the product stories in book *The Black Swan; The Impact of the Highly Improbable* by N.N. Taleb, we learned about a lot of unforeseen successes of new products that the designers did not expect. On the other hand a lot of marketing studies and details of the new product characteristics are not enough to ensure success for a new product. An example of a brand new product today is the electronic book (with e-ink). A lot of producers have developed one but the big bang for this type of product has not yet arrived.

The Kano model teaches us about the great influence of delighters in making a product a successful one, but foreseeing what a delighter will be can only be achieved after the introduction. So designing a brand new product will always entail a great amount of uncertainty for each design team. On the other hand a lot of product development will be done on existing products, which will mean improvements to existing products, bringing in new wishes from clients, or thinking of some surprises in the form of delighters, etc.

In this chapter we will start the product development process from the point of having a form of product description or product definition document in hand. We will concentrate on the factors and of all the phases of the Life Cycle that are encountered in designing or developing a (new) product.

5.2   Designing over the Whole Life Cycle

Under Life Cycle we will consider all the phases a product will face during its total life time. That means in general the following phases:

- ideation or conception phase;
- design phase;
- engineering phase;
- manufacturing phase;
- installation and commissioning phase;
- operational phase;
- end of life or out of operating phase.

Each phase will be considered and allocated their place in the integrated design process. Some phases take only days or weeks and other phases may take years. Central to this business process is value creating, so we will consider the value creation over the whole Life Cycle and will identify in each phase the benefits for both the maker and the client.

Designing over the Life Cycle as a business process is executed by a multidisciplinary design team, which will be composed of all the stakeholders involved with the product (functions, structure and architecture). For more details about the organizing and working of this design team see chapter 2.

First we will go through the defined phases and also give the criteria for each phase that can be applied to achieve successful results. Each phase will end with a formal document. So it is possible to present the results in a way that the design team can easily apply the concurrent engineering methodology.

5.3   Phases over the Life Cycle

5.3.1 Ideation or conception phase

The result of this, essentially, pre-phase of the Life Cycle is the 'PDF Document' (PDFDoc), as we have seen in chapter 4, which includes all the client wishes about the (new) product. In figure 5.1 we see an example of the different possible phases over the Life Cycle for a development process of a chemical production plant, with feasibility and assessment as important decision points for go / no go decisions.

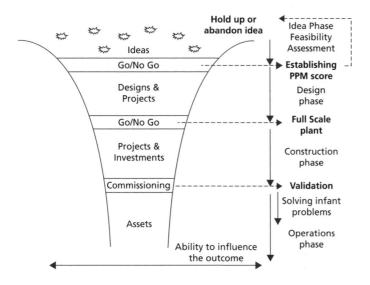

Figure 5.1 Design Process over the Life Cycle

5.3.2 Design phase

In this phase the outcomes of the PDF Document will be used as the starting point for the design process. Here the main translation of client wishes into a real product will take place. Firstly the next chambers of the House of Quality will be filled and a complete functional decomposition model of the proposed product (a Hamburger model) will be constructed. In this model all the possible solutions to formulate functions on all levels are shown. The model will be also applied for all the administrative affairs generated by the integrated design process. Everyone has to realize that administration is a big part of design work. How to store the data in a structured way is one of the main topics of IDE!

Further investigations are started about the possible solutions at all levels of the decomposition through, for example, the TRIZ methodology, a study of relevant patents or of the solutions of competitors. It is important to look at all the possible solutions of those functions, which may have more than one possible solution, or are still completely unknown to the design team. Also knowledge of the applied technology of competitors has to be investigated in this phase.

During the design phase all possible techniques or methodologies of design can be applied (a lot of design books for this phase are available). Solutions are formulated for each function in the product and the design aspects are given, along with RAMSHE specifications and material characteristics. The possibilities of failure for each function have to be investigated, including the criticality of failures of each function in relation to main functions. During this phase, cost estimates of all function-solution combinations are developed, so an indication of a possible cost price can be given. If needed, a Life Cycle cost price is also estimated, particularly when more than one solution for a function is possible!

The outcome of this phase is the **Product Design Document (PDSDoc)** in which the functions, the functional specifications and the possible solutions are given for all levels. This means in general the next levels of the decomposition are recognized: main function-product level, module level, assembly level and part level.

Special attention has to be paid to those functions with more than one solution, where the final situation still is unclear, in order to consider which solution is the best. In the next phase the final choice has to be taken.

5.3.3 Engineering phase

With the input of the PDSDoc, the engineering work in this phase will comprise the investigation and description of all the specifications for parts, assemblies and modules of a product. Furthermore the drawings of parts, assemblies and modules, and 3D models of each are made. When there is more than one solution possible for a function (on each level), a selection must be made during this phase, because a lot of important aspects of further phases are brought in during this phase. If this is still not possible, then two or even more prototypes will be in development until testing is carried out in the manufacturing phase.

This phase considers a lot of aspects of the client wishes and the Life Cycle in relation to the product to be developed. Typical aspects with a high level of attention are:

- visualisation of client wishes;
- sustainability, pollution and energy;
- RAMSHE specifications and ideas about robust design (Taguschi);
- cost price and Life Cycle costs;
- design issues and the real user (operator).

Input from RAMSHE specifications
It can be very useful to formulate at this stage the so called RAMSHE specifications. These stand for:

- R = Reliability;
- A = Availability;

- M = Maintainability;
- S = Safety;
- H = Health;
- E = Environment.

Example

A manufacturer of electrical transformers has to fulfill the following RAM specifications:

$R \geq 0.99$, so the amount of failures has to be $\lambda \leq 5.95 \cdot 10^{-5}$ per 168 hours.

$A \geq 99.6\%$, every week 15 minutes for maintenance, every one day for inspection and oil control and or change.

Maintenance activities have to be done in the aforementioned 15 minutes a week and one day a year.

At each functional level these specifications have to be documented (for modules, sub modules and parts). See also chapter 7.

Input of manufacturing, installation and operation, and maintenance

During this engineering phase the work provides a strong input to the important issues of manufacturing, installation and maintenance, so that all these aspects are considered and are incorporated into the parts, assemblies, modules and product. Also the decision of making, buying or subcontracting of parts, assemblies and modules will take place during this phase.

Typical manufacturing questions are:

- standard or individual designed parts (assemblies, modules);
- choice of whether to make or buy parts (assemblies, modules);
- easy to make;
- easy to assemble;
- easy to test.

Typical installation questions are:

- easy to transport (container dimensions);
- easy to install (without help of external means);
- easy to commission (so a client has a good feeling about the final result).

Typical operation and maintenance questions are:

- easy to operate (all the operational functions are easily placed for the operators);
- easy to execute the daily routine maintenance and inspection tasks;
- easily changeable worn parts.

The outcome of this phase is the **Product Engineering Document (PENDoc)** which contains, for each part, the assembly, module and final product specifications of how to manufacture, assemble, test, install, commission, operate and maintain. All the drawings for production, assembly, installation and maintenance are included, together with a complete 3D model of the product. The PENDoc has to provide a clear translation from the client wishes to an operable product in all aspects.

5.3.4 Manufacturing phase

The manufacturing phase is again divided into three sub phases: making of parts, assemblies and modules, assembling all together into the final product; and the final test. Each sub phase can sometimes be partly divided in other sub phases. Shortly each type of phase will pass our review. The input of the PDSDoc and PENDoc is clear at this stage.

Making of parts, assemblies and modules
Depending on the choices made during the engineering phase, the parts, assemblies and modules are either made by the company itself or bought outside. Outside products can be standard (e.g. bearings, bolts and nuts, etc.) or made by other manufacturers (the suppliers). All the parts, assemblies and modules that need to be made are controlled by the specifications of the **Product Manufacturing Document (PMADoc)**. So during the engineering phase the production department has to deliver the correct inputs for the manufacturing options of the parts, etc. This means that during this stage faults in drawings etc. must be minimal. During the manufacturing period proposals for improvement will be brought to the table. The production team has to collect these proposals and present them to the design team for product improvements.

Final assembly
The ways to construct parts into assemblies, assemblies to modules and modules to the final product are also controlled by the PMADoc document. Once again, the input of the production department is very important to this document. In addition, the process of assembling has to be the subject of improvement studies, because manufacturing and assembling has a great influence on the cost price of a product. The improvement proposals will again be brought to the table of the design team.

Testing and commissioning

After the final assembly the testing phase starts, and the test procedure often takes place in conjunction with a client. During the final test the product has to show that it fulfills all the specifications from the PDFDoc document. The tests on parts, assemblies and modules take place earlier in the assembly process, mostly at the final stage of manufacturing of these parts, etc. Again all these are fully described in the PENDoc document. The commissioning of the product by the client can take place at the end of a successful final test, or at the final installation in a production plant. Models such as the Hamburger model or the V-model can be very helpful for the set up of a testing program.

The final product will be guided by a final product document PENDoc in which is described the final state of the (commissioned) product that will leave the factory, with all the part numbers, assembly drawings, installation survey and drawings and test results,

5.3.5 Installation phase

The installation phase contains some sub phases, the installation phase, the start up phase and the final commissioning phase. The product should be developed and designed in a way that ensures that the construction on site does not cause major problems. This means well designed parts, assemblies and modules (for big products), well designed help instruments for putting together large components and clear installation drawings and instructions. After the construction of the product (or installation) the starting up phase starts. This often begins with cleaning activities, including the piping systems. Sometimes these piping systems cause major problems because of 'rust' in the pipes, which can result in significant pollution troubles in e.g. lubrication oil systems. A long time spent on effective cleaning is therefore frequently required.

After erection and cleaning, the start up of the machine or installation can take place and the final tests can be carried out. After these test runs have been carried out successfully, the final commission or take over of the product can be executed.

The end of this phase will be completed by a document referred to as the **Product Support Document (PSPDoc)**. This document provides the installation instructions (installation manual) and includes all the data of the installation phase, including the test results.

5.3.6 Operational phase

After the final commissioning, the real life of the product (machine, installation) starts. Guided by the operation manual part of the PSPDoc, the operation can be started. The operational manual will typically consist of two or three parts: operation, maintenance and improvement. In the operation part the normal start and stop procedures will be given and operators use this part for training and daily work. The maintenance part gives an overview of all the normal wearable parts which can be changed at the right time in preventive maintenance schemes. In chapter 7 a methodology for developing a maintenance plan will be given. The maintenance plan is based on a functional decomposition with the Hamburger model. With this decomposition model it is possible to develop an overview on all levels of the functions and the functional specifications. The input for all maintenance or repair activities will be the failure of a function, which means a function does not fulfill the functional specifications and is in a state of functional failure.

All machinery will be subject of improvements. So it can be very fruitful for a manufacturer to form a group of clients (users) which are pro active in passing back ideas about extra or new features and for improvements in weak parts. In the world of internet the name **crowd sourcing** is introduced to give this phenomena a title. A lot of companies apply this method for gaining new ideas (see also chapter 4), for looking at possible improvements and for identifying potential delighters.

The end of this phase will be a finalized Product Support Document (PSPDoc). This document includes all the factors relating to commissioning, installation, operation and maintenance.

5.3.7 End of life phase

After a hopefully fruitful life of operating, the product will come to the end of life or **out of function phase**. Dismantling takes place and the potential recycling of materials needs to be considered. For the design team this will mean that during the design and engineering phase all parts, assemblies and modules have to be specified bearing in mind sustainability issues, so that it can be decided what will happen to each part at the end of life phase. Is it reusable, repairable, can the material itself be reused again, etc. This means that the design team must pay attention to this aspect during the specification stages.

Note: Materials used for of car production nowadays are more than 98% reusable or recyclable.

5.4 Concurrent Engineering Planning

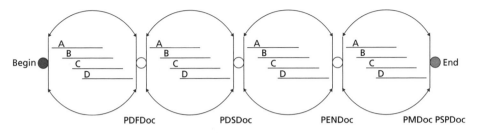

Figure 5.2 Concurrent Engineering Planning (2)

In this **Concurrent Engineering Planning model** we can easily mark the points on the time scale when the documents have to be ready. But work can also be carried out concurrently on the different reports after the PDFDoc is finalized and the important go / no go decision has taken place.

5.4.1 Overview of documents over the Life Cycle

Over the development time of a product, the product development team produce a document at the end of each phase to store all the data and decisions made during the process, This is not only important for starting the next phase but also for future improvement activities carried out during the operational life time of a product (see chapter 6), where it is very helpful to reconstruct the original decisions and data. Of course this information has to be stored in a structured way (e.g. Hamburger model and codes) and be easily accessible.

5.4.2 Product development team over the Life Cycle

In figure 5.2 we can see very clearly that the scope of the product development team during the development process changes from pure client wishes and product definition to a real product that is easy to make, to operate and to maintain. So, the composition of the team can also change.

The design team that is put together for designing and specifying the new product will, in addition to having representatives from the world of the client, also have representatives from the phases of manufacturing, installation and commissioning, operation and maintenance, and end of life. All the aspects they bring to the design process have to be weighed and evaluated on basis of conscience and equality. In order to enhance the process of bringing in and evaluating all these aspects over the full Life Cycle the team members have to apply the methodology of concurrent engineering. In chapter 2 the design team composition aspects have already been addressed.

5.5 ICT over the Life Cycle

The development of a product over the Life Cycle needs an information structure based on structure models like the Hamburger model. We also argued in chapter 3 for the function of a 'structurist' who would be responsible for safeguarding the right structures and codes in the engineering databases for the effective storing and accessing of data and information. In figure 5.3 we see an example of this database (the so called WEBIM model) especially for the execution plain of the IDE company model (see figure 1.7 or figure 4.2), where the processes from quotation to maintenance take place for sold products.

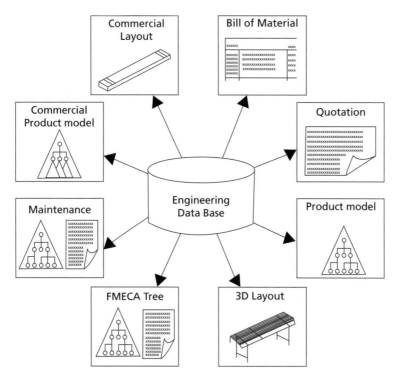

Figure 5.3 The WEBIM model

We see that over nearly all the phases of the life cycle, with the exception of the assembly part of the process product planning, there are other ICT modules like Enterprise Resource Planning (ERP) running the process. Also here we can apply the structural model of the Hamburger decomposition model. By turning the model 90 degrees to the right (figure 5.4) we see on the right side of the picture the final product (that fulfills all of the desired functions) and from left to right on this picture we can see the parts, sub assemblies, assemblies, modules and the final product. When we put a time line under the horizontally oriented Hamburger model we can make a **planning model** for manufacturing and final assembly.

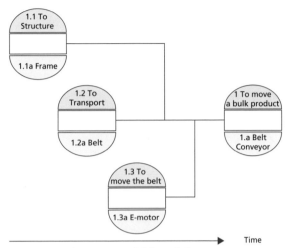

Figure 5.4 Planning model for a belt conveyer

To summarize: control over good knowledge stored in a well organized database brings about a lot of potential profits for a company in the form of quicker access to perfect data, faultless products of perfect quality and perfect planning, a lot of profit in time (between departments) and a reduction or elimination of potential claims (of clients).

5.6   Costs of Manufacturing, Life Cycle and Investment

One of the drivers for a successful product is the costs. Not only the manufacturing costs or the cost price, but also the typical operational costs like Life Cycle Costs and investment costs. The development or design team must continually calculate these costs during the development of the product. And search for alternatives for functions and/or solutions, when the costs become too high for a profitable operation, or in order to compete with other manufacturers. A cost estimator has to be member of the team. When making the different type of calculations, the Hamburger model can be applied to gain an insight and overview of the costs.

5.6.1 Manufacturing costs

Manufacturing costs in general consist of material costs, labour costs, machine costs, stock costs and control costs. During the development of a product these costs must be estimated for each function on each level. When there is more than one solution for a function, the costs for each solution will be estimated. The costs can be placed beneath each Hamburger in the model and this provides an excellent overview of the costs on each level (see figure 5.5).

Note: The sum of the sub functions in figure 5.5 is €910,-, but for assembling them together we calculate 5% assembling costs.

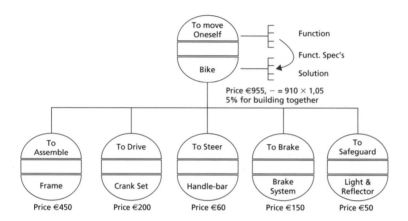

Figure 5.5 Hamburger model and costs

The development team will be continuously confronted with the cost implications of their decisions. In situations where there are many solutions or strong competition in the market, then the manufacturing costs are of immense importance.

5.6.2 Life Cycle Costs

In addition to the manufacturing costs, other costs that are of particular interest to clients are the yearly operational costs and the **Life Cycle Costs (LCC)** over a period of years.

Typical yearly operational costs are: raw materials to process, energy costs, pollution costs, labour costs and maintenance costs. To compare the different solutions from each other, the energy, pollution, labour and maintenance costs are the most important.

Life Cycle Costs consist of investment costs, operational costs over 'n' years and the out of service (end of life costs). The investment costs are not only the selling price but are all the costs required to make the product ready to operate and are, in general, the sum of selling price, installation and commissioning costs, training costs, and the costs of the first set of maintenance spare parts.

In schedule the Life Cycle Costs for product A and for product B are:

LCC	product A	product B
• Investment costs (+ 1 time)	€55000	€50000
• Operational costs (+ n times) = 10 years		
• Product A: €5000,-/yr	€50000	
• Product B: €5750,-/yr		€57500
• End of life costs (+/- 1 time)		
• + end value (e.g. car) = 10% investment costs	€ – 5500	€ – 5000
• – demolish costs (e.g. factory)		
Total LCC for A and for B:	€99500	€102500

So product B is cheaper at the moment of investment but is more expensive when viewed over the whole Life Cycle.

During the development of a product the LCC calculations can be very helpful in investigating the different solutions for a given function, or to compare the organization's own solution with that of a competitor. It gives an insight in the consequences of the costs over a long period of time (10 - 20 years). It will always be used as a comparable calculation between at least two and more possibilities.

5.6.3 Investment costs calculation

Another method used to compare the costs of two of more solution of competitive products is the **investment costs calculation**. This is especially the case when it is not all about costs but also when income can be generated by the product. The most common calculation and also the most relatively simple is the calculation of the **Pay Back Period** (PBP, mostly time in years). This is the time that it takes to earn enough income so that all the investment costs are paid. In formula form this looks like:

Pay Back Period = Investment costs / Income per year

Example of a Pay Back Period calculation, again for the products A and B:

PBP calculation	product A	product B
• Investment costs	€55000	€50000
• Income per year	€30000/yr	€27500/yr
Pay Back Period is:	1.83 years	1.82 years

Notes:

For the LCC calculation product A gives the best result, whilst with the PBP calculation product B is a little bit better than A. In case of small differences the design team can decide that other factors will be used to make the choice between A and B or can decide to calculate in a heavier way. For example this could be a calculation of the Net to Present Value, or the internal discount rate.

Also the investment cost calculations are applied for comparing different alternative solutions. This is especially the case in business investments. Each competitor on the market must have an idea about the strengths of the others in the field of profitable investments.

5.7   Parts Engineering

After finalizing the investment decision about the functional desires on each level and studying the possible solutions on each functional level, the more or less detailed engineering of modules, sub modules and parts comes in the picture. Again the Hamburger model is very useful at this stage of the engineering work because all the decisions about modules, sub modules and parts can be illustrated in this model in a very easy to assimilate way, e.g. by colors, from the color green for clear choices through to the color red for great uncertainties . The design team can make a clear plan by looking at these uncertainties and can also make fair estimates about the planning times needed to finish the design work on the red solutions. Also is it possible where there are uncertainties about the right solution to design concurrently two or even more solutions. A lot of time and money can be gained by adopting this procedure during this stage rather than after the building or prototype or testing stage! Because then a completely new solution has to be designed, built and tested again.

In addition to the Hamburger model, there are two other models that are very helpful. We will briefly cover the FMEA (Failure Mode and Effect Analysis) and the FTA (Fault Tree Analysis) models.

Beside these models it is also desirable to formulate the so called RAMSHE specifications (see also § 5.3.3) for each functional level as part of the design criteria for solution, modules, sub modules and parts.

5.7.1   FMEA model

FMEA stands for **Failure Modes and Effects Analysis** and gives an overview of all possible failure modes and their effects on the main function. This model will also be applied at the maintenance phase (see chapter 7) in the form of a Functional critical FMEA (*Fc*-FMEA).

When applying this model we will start by making a decomposition of a system. On each level we can ask to the sub systems, parts, etc. what failure modes are possible and what will be the effect of each failure mode.

Example

To show the first principles of the FMEA we apply it to a system that is a simple electric circuit with a lamp, a battery, a switch and connected by cables (see figure 5.6).

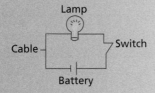

Figure 5.6 Simple electric circuit

Decomposition in functions and sub functions gives:

Function: *light on demand*   Solution: *electrical light system*

Sub Functions: make light   lamp
                 generate power   battery
                 control   switch
                 transport power   cables

Now we develop an overview of all possible failure modes and their effects on the system (see table 5.1).

Solution	Failure Mode	Failure Effect	Comment
1 Lamp	1.1 Open filament 1.2 Shorted base	1.1 No light 1.2a No light 1.2b Fire hazard	System fails System fails System destroys
2. Battery	2.1 Low charge 2.2 No charge 2.3 Over charge (Volts) 2.4 No contact	2.1 Dim lamp 2.2 No light 2.3 Lamp blow up 2.4 No light	Go to No charge System fails System fails, damage lamp by over-current System fails
3. Switch	3.1.Fails switch open 3.2 Fails switch closed	3.1 No light 3.2 None, light stays on	System fails Battery to low charge
4. Cable (Filament)	4.1 Cable broken	4.1 No light	System fails

Table 5.1 Simple FMEA

5.7.2 *Fc*-FMEA

In this book we develop a special variant of the FMEA model, the so called **Fc-FMEA**, based on the criticality of a functional failure of a (sub)system. In chapter 4 we discussed all the client wishes and translated them in a whole package of client wishes in the PDF Document. With the input of this document we can compose the *Fc*-FMEA. We start with the complete functional decomposition of the product in the form of a Hamburger model and bring the specifications to the different systems, where they belong. On each level of systems each specification can be classified as very critical, critical, less critical and not critical. The **criticality** is defined here as the capability of the function to dismiss the main system by failure of this function (or not able to fulfill the functional specification). By giving each system a color that indicates the level of criticality (e.g.: red for very critical, orange for critical, yellow for less and white for not) it can made very easy visible in the Hamburger model which are the dangerous systems and which systems do not expect any trouble.

If the system (of each level) is unable to fulfill the functional specification(s) then this system is in state of functional failure and a repair or restoration action will need to take place. For the design team it is very useful to define in a Hamburger model the functional specifications and the associated functional failures. By giving these functions and functional failures a possible criticality and making a picture with colored functions it is also possible to develop a complete overview of all 'dangerous' or 'red' functions in a system. The designer has to pay special attention to these functions in order to decide what to do. E.g. redesign, double solutions, etc.

When no real failure data is available, a **criticality matrix** can be used in which each function can be marked. In this matrix we see on the vertical the frequency F and on the horizontal the severity S (see figure 5.7). The team members have to give their ideas on how critical each of the functions are.

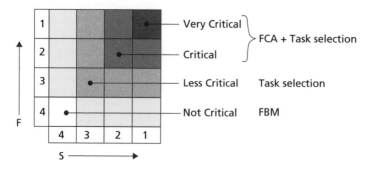

Figure 5.7 Criticality matrix

Note: Diverse aspects can be made visible in the Hamburger model and each member of the design team can input their own thoughts on aspects of criticality.

Example

Function: 'to make light'
Functional specifications are:

- make light of minimal X Lux;
- can burn for at least Y hours.

Solution: Electrical power lamp

The *Fc*-FMEA can be presented as follows (see table 5.2):

FU	FV	FS	FF	KFF	Part	FM	FE
1	A	1	1	1	1	1	1
Make light	El lamp set	> X Lux	< X Lux	Less Crit	Battery	Low charge battery	Weaker Light
		2	2	2		2	
		> Y hours	< Y hours	Less Crit.		Low charge battery	Weaker light
			3	3		3	2
			No light	**Critical**		Battery empty	No light
					2	4	
					Cable	Cable broken	No light
					3	5	
					Switch	Switch open	No light
						6	3
						Switch closed	Later no light Battery empty
					4	7	2
					Lamp	Lamp filament broken	No light

Table 5.2 Simple Fc-FMEA

With the *Fc*-FMEA (see table 5.2) we can develop an overview of all weak or critical functions in a system, so we can incorporate these into the design stage measurement in order to improve the criticality of these functions. In the overview we can mark the critical and very critical functions by colors.

5.7.3 Fault Tree Analysis (FTA) model

FTA stands for **Fault Tree Analysis** and it can analyze for each single failure the possible causes. It can also be applied for maintenance failure analysis (see chapter 7). To develop a FTA-diagram we will use the same simple electrical circuit of the FMEA (see figure 5.8.).

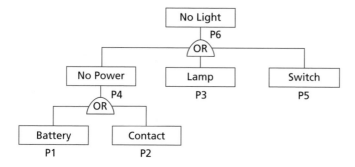

Figure 5.8 FTA (for light circuit)

The main fault of the system is 'No Light' (from the lamp), this can be as a result of four possible failures: Or by Lamp failure, Or by No Power, Or by Switch failure, Or by broken Cable.

The No Power failure can be as a result of two failures: Or Battery failure, Or Contact failure. We see in this relatively simple scheme only the so called ´OR´ gate. Mostly the FTA scheme has a combination of ´OR´ and ´AND´ gates. When we know the probabilities of each failure it is possible to calculate the final probability of the main failure (No Light).

Example

Simple electric circuit

The following probabilities of failure are assumed:

Battery failure $P_1 = 0.01$, Contact Failure $P_2 = 0.01$, Lamp failure $P_3 = 0.001$, No Power failure P_4 (to calculate), Switch failure $P_5 = 0.02$ and No light failure P_6 (to calculate)

Calculation (following the FTA tree (see figure 5.8).

$P_4 = P_1 + P_2 - (P_1 \times P_2) = 0.01 + 0.01 - 0.01 \times 0.01 = 0.0199$

$P_6 = P_3 + P_4 + P_5 - (P_3 \times P_4) - (P_3 \times P_5) - (P_4 \times P_5) + (P_3 \times P_4 \times P_5) = 0.001 + 0.0199 + 0.02 - (0.001 \times 0.0199) - (0.001 \times 0,02) - (0.0199 \times 0,02) + (0.001 \times 0.0199 \times 0.002) = 0.04046$

5.8   Design Team and Part Engineering

Part Engineering means looking at a lot of possible solutions for the functions at each level. Several methodologies are available to do this work of selection in a proper way. We will provide a brief overview of some of the most important methodologies on this field:

- TRIZ; Systematic Innovation, J. Terninko a.o.;
- Don Clausing methodology, EQFD;
- Pugh methology;
- Van den Kroonenberg methodology;
- Taguchi methodology;
- Simple Matrix for Selection.

We will also give some simple but very useful tools that the design team can apply for doing this kind of design work.

Furthermore it is advisable to apply the Hamburger model for administrative reasons, in order to store all the data about decisions the team will make about a range of solutions: techniques, technologies, materials, costs of manufacturing, Life Cycle Costs, RAMSHE spec's, etc.

The design team has to decide which methodology or combinations of these methods will be applied during the part engineering stage.

5.8.1 TRIZ, a system of methods

TRIZ is a combination of methods, tools and a way of thinking, with the ultimate goal being to achieve absolute excellence in design and innovation. The TRIZ key elements are:

- *Ideality.* Ideality is the ultimate criterion for system excellence; this criterion is the maximization of the benefits provided by the system and minimization of the harmful effects and costs associated with the system. The definition of Ideality is:

Ideality = $\sum$ benefits / ($\sum$ costs + $\sum$ harm)

So the development of any technological system will be always towards the direction of:
- increasing benefits;
- decreasing costs;
- decreasing harm.
- *Functionality.* Functionality is the functional building block of system analysis. It is used to build models showing how a system works as well as how a system creates benefits, harmful effects and costs.
- *Resource.* Maximum utilization of resource is one of the keys used to achieve maximum Ideality.
- *Contradiction.* Contradiction is a common inhibitor for increasing functionality; removing contradiction usually greatly increases the functionality and raises the system to a totally new performance level.
- *Evolution.* The evolution trend of the development of technological systems is highly predictable and can be used to guide further development.

Based on these five key elements TRIZ developed a system of methods, a complete problem definition and solving process. It is a four-step process consisting of the following steps:

I   Problem definition
This is a very important step in TRIZ. If the team defines the right problem and does it accurately, then that will be 90% of the solution. The problem definition step includes the following tasks:

- *Project definition.*
- *Functional analysis.* This includes the function modelling of the system and analysis. This is the most important task in the 'definition' step.
- *Technological evolution analysis.* This step looks into the relative maturity in technology development of all sub systems and parts.

- *Ideal final result.* It may never be achieved but this provides us with an 'ultimate dream' and will help us to think 'out of the box'.

2   Problem classification and tool selection
TRIZ contains a wide array of tools for inventive problem solving. Firstly we have to classify the problem type and then select the tools.

3   Solution generation
TRIZ tools are available to generate solutions to the problem.

4   Evaluation
In any engineering project the soundness of the new solution needs to be evaluated.

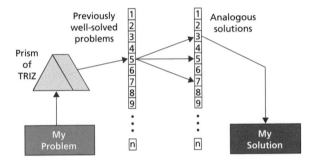

Figure 5.9 TRIZ Approach to Problem Solving

TRIZ is the acronym for Theory of Inventive Problem Solving. The TRIZ methodology originates from the former Soviet Union and was created by G.S. Altshuller. The basic idea behind the methodology is the statement that nearly every possible solution of a function is already described (see figure 5.9) in patents literature. By studying this literature, initially reviewing over 200. 000 patent abstracts, he discovered five levels of innovation:

- Level 1: Apparent or conventional solution 32%; solution by methods well know within a specialty.
- For example: level 1 invention is the increase of the wall thickness for greater insulation in homes.
- Level 2: Small invention inside paradigm 45%; improvement of an existing system, usually with some compromise..
- For example: existing system is slightly changed and includes new features that lead to definite improvements.

- Level 3: Substantial invention inside technology 18%; essential improvement of an existing system.
- For example: replacement of manual standard transmission with an automatic transmission in a car.
- Level 4: Invention outside technology 4%; new generation of design using science not technology.
- For example: materials with thermal memory.
- Level 5: Discovery 1%; major discovery and new science
- For example: the laser.

Notes:
About 95% of innovations are more or less further developments of existing or already known systems (sum of levels 1, 2 and 3)!

The levels 4 and 5 we have termed in this book as ´risky´ because the real potential of the products in relation to the markets is unknown.

The TRIZ methodology can be usefully applied for those situations where a design team want to consider completely new ways or ideas in their design. A lot of new ideas for new solutions of formulated functions can be generated.

5.8.2 Don Clausing methodology for module and part selection

Clausing has developed an enhanced methodology known as **Enhanced Quality Function Deployment (EQFD)** from the QFD methodology, in which he describes a complete approach from specifying client wishes (Voice of Customer) through to product definition and module and part design, via product planning, design, process planning and operations planning. He has developed a model with a series of Houses of Quality, where the output of one HoQ is the input for the next (see figure 5.10). For each stage in the process a document has to be produced for collecting and storing all relevant data.

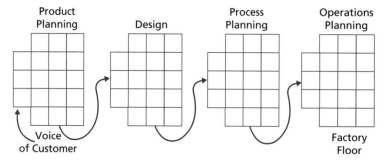

Figure 5.10 The EQFD methodology

His design methodology follows the Life Cycle and works also with the distinction between functions and solutions.

For successful engineering work he lists the following important enablers as part of this process:

- the concurrent process (for the design team);
- focus on quality, cost and delivery (times on schedule);
- emphasis on customer satisfaction;
- emphasis on competitive benchmarking.

5.8.3 Pugh methodology for concept selection

Pugh has developed a **concept selection matrix** for the total system architecture, sub system and parts concepts. We start by defining the so called 'Datum' or reference concept with the criteria to consider. These criteria are based on the outcomes of the House of Quality (see PDFDoc). The selection matrix consists of columns for criteria, concepts and Datum, and rows for the criteria specifications of the Datum and concepts to consider (see figure 5.11).

The team compares the criteria of all the concepts to that of the Datum. When a concept promises a better ability to satisfy for criteria A, then it is given a 'Plus' mark, for equal it is given an 'S' mark, and for inferior a 'Minus' mark. So each criteria of the Datum (A, B, C, etc) will be handled for each concept (see figure 5.12). At the end it is possible that one concept is far better than the others and can become the new Datum, or a new Datum is constructed by the best outcomes for each criteria or from combinations of concepts. The progress of generating new concepts and comparing these with the Datum can be continued until the team is satisfied or when no other concepts can be found. In principal it can be a continuous progress.

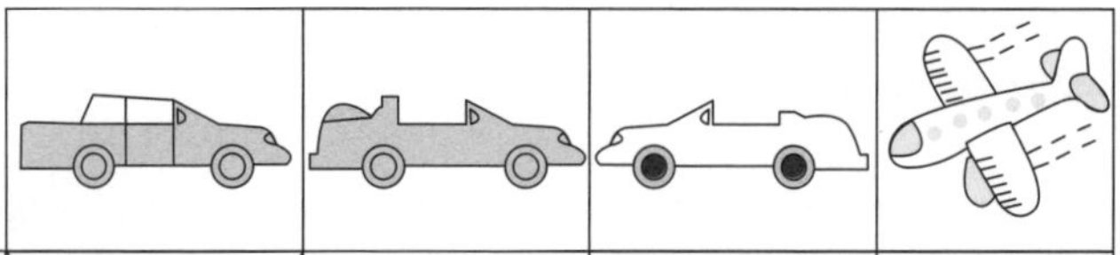

DATUM			
Response: quicker is better		+	+	+
User maintenance: fewer tasks is better		−	−	−
Ease of mfg.: fewer parts is better		S	−	−
Ease of service: clearer messages is better		+	+	−
Risk: older technology is better		−	−	−
Mfg. risk: more off-the-shelf parts is better		S	−	−
		+ 2	+ 2	+ 1
		S 2	S 0	S 0
		− 2	− 4	− 5

Figure 5.11 The Datum concept

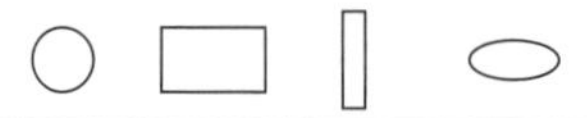

Concepts				
Criteria	○	▭	▯	⬭
A	+	−	+	D
				A
B	+	S	−	T
				U
C	−	+	−	M

Figure 5.12 The Pugh methodology

5.8.4 Van den Kroonenberg methodology for concept selection

Van den Kroonenberg provides a concept for generating possible solutions by selecting or searching for solutions within each function (all levels) and combining these partial solutions into one or more total concept solutions which can then be further evaluated. With the help of this so called **Morphological Scheme** (see figure 5.13) it is possible to select a technical solution for fulfillling a function. In this type of scheme all possible solutions are presented and from these a solution is chosen. In his methodology weighting matrices can also be applied (see figure 5.14 for an example). One matrix is for the client wishes or user demands and the other for manufacturing demands.

Figure 5.13 Morphological Scheme

function description	working methods									
	1	2	3	4	5	6	7	8	9	10
To deliver energy for transport	man	e-motor	combustion engine	diesel powered engine	solar energy	animal power	m.g gravity			
Guided transport	hand	rail	wheels	roller conveyor	air glider	telpher carrier/ trolley chain				
Positioning flat plate	hand	Lift	pulley	hoist	monorail	hydraulic	spindle			

Figure 5.13 Part of a Morphological Scheme

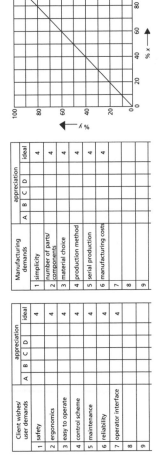

Client wishes/ user demands	appreciation				
	A	B	C	D	ideal
1 safety					4
2 ergonomics					4
3 easy to operate					4
4 control scheme					4
5 maintenance					4
6 reliability					4
7 operator interface					4
8					
9					

Manufacturing demands	appreciation				
	A	B	C	D	ideal
1 simplicity					4
2 number of parts/ components					4
3 material choice					4
4 production method					4
5 serial production					4
6 manufacturing costs					4
7					
8					
9					

Figure 5.14 Example of (a part of) a Decision Matrixes (in this case machine automation)

5.8.5 Taguchi methodology for robustness

Taguchi has developed a series of methodologies around robustness. He describes **robustness** for the design of a product, the stability of the production process and the reliability behavior during the use or operating phase of a product. For each of these it is possible to formulate possible instabilities, called 'noises'. The design team has to consider these noises in order to eliminate those which give rise to unstable situations. E.g. if the design contains parts with a potential wide variation in outcome of physical properties which also result in unstable behavior of the product, then the noises which generate these instabilities have to be eliminated. It is the same for production processes, where parts with wide variations in measures can lead to troubles at the assembly stage. During the operational stage the robustness measures the degradation behavior of a product or a part, expressed by the so called **Mean Time Between Failures (MTBF)**, with a high mean time equating to a stable operation.

Taguchi's philosophy has far reaching consequences, yet it is founded on three very simple and fundamental concepts, these concepts are:
1. Quality should be designed into the product and not inspected into it.
2. Quality is best achieved by minimizing the deviation from a target. The product should be so designed that it is immune to uncontrollable environmental factors.
3. The cost of quality should be measured as a function of deviation from the standard and that the losses should be measured system-wide.

These three concepts were his guides in developing the systems, testing the factors affecting quality production and specifying product parameters.

Notes:
- The famous Deming had already observed that 85% of poor quality is attributable to the manufacturing process and only 15% to the worker.
- The IDE philosophy states that in the design of a product not only the cost price but even the future profit is included, so that during the product development process concurrently the right manufacturing methodology has to be developed.

5.8.6 Selection matrix

A relatively simple methodology for the selection of systems, subsystems or parts is the **selection matrix**. Here, formulated requirements, wishes and specifications are split into two sections: *requirements* and *wishes*.

For each requirement only the final result counts, this has to be fulfillled and the answer is only 'yes' or 'no' (fulfillled or not fulfillled)

For each wish the key factors are how it relates to the outcomes of competitive solutions. By weighting these factors the relative weight of a solution can be compared to another. The two weighing factors are a percentage of 100% (the solution that fulfills the requirement most effectively becomes the highest value, than the second etc) and a multiplayer from 1 to 4 for each wish, so we can also differentiate between the wishes. For each solution a wish is calculated by multiplying the percentage and the multiplayer. The sum of these calculations gives the end score for each solution.

One potential problem is defining what is really a requirement and what is a wish. Sometimes the difference between these is very small

The winning solution is the one that has only 'yes's' for the requirement and the highest weighting score.

Example

For the selection of a part the requirements are:

- costs;
- material used;
- robust to produce.

The wishes are:

- weight;
- easy to transport;
- robust to operate;
- easy to change.

Four solutions are identified out of the first selection as possible candidates, and an analysis carried out using the selection matrix gives the following result:

Possible alternative solutions:	A	B	C	D
The requirements are:				
• Costs	+	−	+	−
• Material used	+	+	+	+
• Robust to produce	+	−	+	+

The solutions A and C fulfill all the requirements with + (yes) and are therefore investigated further:

The wishes are:
- Weight: 1 × (100)   1 × 30 = 30   1 × 70 = 70
- Easy to transport: 2 × (100)   2 × 60 = 120   2 × 40 = 80
- Robust to operate: 3 × (100)   3 × 70 = 210   3 × 30 = 90
- Easy to change: 2 × (100)   2 × 45 = 90   2 × 55 = 110

Total: = 450   = 350

Result: solution A gives the best outcome and can be chosen for the application

5.9   Solutions for Modules and Parts

During the decomposition process we determine the main function with sub functions, sub sub functions, etc. with, as counterparts, the main solution, the modules, the sub modules, assemblies and parts. From the PDSDoc all the wishes of clients are catalogued and translated in functions and possible main solutions. Now we can start the process of selecting the solutions for the sub functions and the sub sub functions. When we look at the next Hamburger model (figure 5.15) than we see that there is more than one solution for a function. We have seen from the TRIZ methodology that only 4% of the new design projects come from scientific research, and only 1% are completely new discoveries, so the main function will generate more than one potential new solution.

An exciting actual example is the wafer step technology for making chips. The existing technology has reached the specific physical light boundaries and does not have much room for making improvements to smaller line patrons on the wafers. What will be the next step for reducing these line patrons? Several potential physical processes are being investigated now, but the winner is uncertain for each manufacturer in the market. To choose the right technology from the potential processes is one of the big battles in design technology now and for each manufacturer of these types of equipment, the most common policy is to carry out research concurrently on all potential solutions.

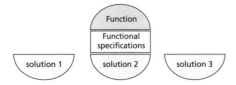

Figure 5.15 Hamburger model 'Function-Solutions'

Example

Bike

Function: `to generate power`

Solutions:

- muscle power by classical crank-chain system;
- muscle power by hydraulic system;
- electrical power by battery.

In most cases the main function has a specific solution and the design team work on solutions that start with the first level of functions, the sub functions. We will further distinguish four levels of solutions: main solution, modules, assemblies and parts (see table 5.3). This is sufficient in nearly all cases.

	Main Function	Main Solution	
Sub Functions			Modules
	Sub Sub functions	Assemblies	
List of Parts			Parts

Table 5.3 For levels of solutions

When a design team wants to review possible new solutions, e.g. hydraulic power against muscle power for driving a bike, then it can apply the TRIZ methodology or carry out a study of patents for possible solutions (see figure 5.16). In this case the team has to look deeper in the possible solutions in order to evaluate the benefits and potential problems of each solution. The uncertainties of totally new solutions have to be investigated, e.g. how to make it, the cost price, operational behavior, reliability, etc.

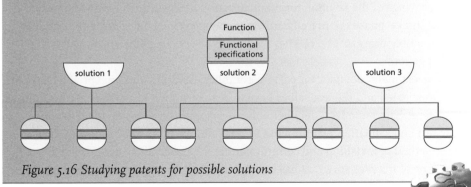

Figure 5.16 Studying patents for possible solutions

In order for the design team to select the right solution each time the following questions have to be addressed:

- what are the functional specifications to be fulfillled?
- can the RAMSHE spec's, when applicable, be fulfillled?
- does the organization have a solution of its own already available (re use of existing knowledge)?
- should the solution be made or bought (with all possibilities to execute these)?
- how will the solution be made?
- is the cost price at an acceptable level (as low as possible)?
- what are the Life Cycle Costs (if applicable)?
- is the robustness of the solution acceptable?

When making the decision about the solution, we have seen that there are a lot of methodologies available:

- Clausing;
- Pugh;
- Van den Kroonenberg;
- Selection Matrix;
- Etc.

During the selection process the design team should pay special attention to Kano's dissatisfiers, satisfiers and delighters. The first two for ensuring they are available and the third for providing the extra's.

At the same time as making decisions about the solutions, the following activities have to be executed:

- Producing the documentation relating to all decisions and choices, so that the existing data can be reused if there is any possible repeat of the process.
- The drawings of the solution for manufacturing and assembly have to be made.
- By the 'make' phase the manufacturing process has to be investigated, including the assembly process. The IDEF-o model can be very useful in carrying out this investigation.
- All the test methodologies for parts, assemblies, modules and final product have to be investigated and the test values have to be determined for all the specifications at each level. The V-model is very helpful to developing an overview of all the tests.
- The test and validation planning has to be set up, including all the test protocols.
- The method of installing and commissioning the product has to be decided.
- The type of maintenance that has be applied in relation to operating the part, assembly, module and product must be decided.

The total package of part selection requires a design team that is composed of all the above mentioned disciplines, which can and has to be work concurrently on these subjects.

5.10 Manufacturing Selection

Manufacturing Selection is also part of the total part selection process, as it has a great influence on the cost price and quality of the developed product. **Design for Assembling** is a well known expression for establishing a mindset whereby, in the design, the question of *'how to produce a product'* has to be solved or provided. When we consider a product, then the question of how to produce it is heavily dependent upon the actual design of the product and the volume of products to be sold (yearly or in total). For example, for cars and household electrical equipment mass production is mostly applied, whilst building a house is mostly done in a small series of activities or single ones. Mass production is executed in specially designed factories with tools especially made for one product, whilst on the other hand generally equipped machine shops can make a lot of different machines which are mostly made on an individual basis. *Mass production* is typically *process driven* with a lot of statistical tools to control it. *Single product production* is typically *project driven*, with critical path methods used as part of a planning model to control the total project time.

The manufacturing selection is also an activity that has to take place concurrently with the part selection process. It starts with the question *'make or buy'*. This question is applicable for each level of functions.

When the decision is *'buy'* than the next decision will be in which form, either via complete outsourcing of a function (with the specifications), or a *'make'* approach to outsourcing on basis of the complete design specifications. With complete outsourcing of a function the subcontractor can become a partner in the business, perhaps investing in the complete development cycle, including the manufacturing, of the function, at their own risk. Also the question has to be considered of do we intend to work with one or with more partners, on a so called *'dual sourcing'* policy. If the *'make'* outsourcing approach is taken to a module, assembly or part on the basis of complete specifications and drawings, then the partner is selected on the basis of manufacturing quality, delivery safety and cost.

When the decision is *'make'*, then the design team has also to consider questions such as: how to make it, which type of machinery and tools are needed, do people need to be trained, is new machinery and/or an assembly shop needed? For example, erecting a new machinery shop takes at least one year and often more than that, so can have a great influence on the launch date for a new product. When completely new technologies for manufacturing come into the picture, then is it worthwhile concurrently considering two or more possibilities to save time.

For all processes to manufacture, the most important items to control for each step are:

- Make or buy and, in case of buy, should this be done by one or more partners (dual sourcing).
- Introducing new manufacturing technologies or using existing proven ones.
- Are capable production staff available and will they have to be trained (in time).
- The cost price to manufacture and the stability of this price in relation to labour and energy costs.
- Material costs, including the stability of these prices in the future.
- Material alternatives in case of unstable delivery, also a form of *'dual sourcing'*.
- Stability of the production process (robustness) in quality and delivery times.

5.11 Design for Manufacturing

In the previous paragraph (Manufacturing selection) we discussed the different possibilities of how to produce a product, including the make or buy decisions. With **Design for Manufacturing**, the main question is how far is a client able to influence the design and engineering process itself. Here we will distinguish four levels of client influence in relation to standard modules and assemblies that the manufacturer has developed to compose the products. For manufacturers a large series of modules and assemblies have a decreasing influence on the cost price, whilst on the other hand a client will distinguish themselves from others by buying exclusive products. The four levels are:

- Complete products, e.g. mass consumer products   100% standard / 0% special
- Assembly to Order Products, e.g. packaging machine: 60% standard / 40% special
- Engineering to Order products, e.g. dredging vessel: 40% standard / 60% special
- Design products, e.g. luxury yachts: 0% standard /100% special

In the case of complete products like consumer products the client influence comes from marketing studies, where the clients' wishes are collected, weighted and prioritized. The manufacturer makes a complete product (in boxes) to sell.

On the other end of the scale is the word of luxury yachts and villas. Here a client discusses with a design team every detail from the hull form of the yacht, to the golden taps in the bathrooms.

Assembly to Order and Engineering to Order are typical forms in the industrial world for making machinery, and in the building industry for school buildings, offices, shops and housing estates.

Design for Manufacturing gives also direction to the way of producing. Mass production is undertaken in very specialist factories that are designed and engineered for one product. Production costs and logistic chains are very important. One the other hand, luxury production takes place in more of a handicraft surrounding with highly specialist craftsman. Creating distinction is more important than costs.

Assembly to Order and Engineering to Order take mostly place in more generally equipped machine shops, with some steps in the process highly automated, e.g. the laser machine for tin plate cutting, and welding robots. These machines often produce the specialist parts of the machines they are selling.

5.12 Design for Operation and Maintainability

During the design and engineering phase of the LCE process, there is a question to be continuously addressed: how will the client operate and maintains the product? So in the design team there should be specialists in operation and maintainability to solve this question.

The important questions for operation and maintainability are:

- How to install a product. For machinery and industrial equipment a good installation manual is needed containing all the practical information about the installation and commissioning of a product.
- How to handle a product. This the world of ergonomics, which means that a product has to be designed in a way that the user/operator can easily find all the handles and buttons, switches, etc. for daily use. The instruments should all be readable in a clear and easy way and also placed on a logical way, etc. The workspace of the user/operator should be optimal and have the right colors and lighting in the working place, etc.
- The design team has to realize that ergonomics provides a lot of chances to distinguish the product from others in relation to Kano's 'delighters'.
- How to maintain a product is also an important factor for client satisfaction. For industrial installations the maintainability is also an important profit and quality aspect. Central to maintainability is the operational behavior during the operational time of a product. If this behavior is stable for long times between failures, then the scores on reliability and availability are high and stable production output figures

are achieved. This means that a product design process which generates modules, assemblies and parts of this quality can fulfill the minimum required values for stable operational behavior, thus the required R and A figures.

Another important factor is accessibility to the parts that have to be regularly changed (the wear and tear parts). Good accessibility means low (work) times for the exchange of parts, short down times for repair and long up times for production. This also means lower costs of repair and higher income as a result of higher production rates.

Beneath pure wear and tear behavior of parts, etc, it is also important to realize that for the real output of a machine the stable production of high volumes of products of good (demanded) quality is perhaps the most important factor for the design team to consider.

5.13 Design for Long Life Service

Besides designing for operation and maintainability, it is also possible to go further and develop a product for long life service. This means that the manufacturer not only delivers the product but also all of the service and maintenance activities during its life time. For the design of the product this mostly means that all the serviceable parts are made in easily changeable modules and that condition monitoring equipment is also applied, even incorporating communication equipment. This enables the service organization to monitor at a distance the behavior or the condition of the product. In fact the product (module) can even warn the service organization of the need for a new part.

For a manufacturer means this type of design provides long life contact with the client. It also provides information as early as possible about the behavior of the product in operation, so that the maintenance process can be monitored continuously, with the financial side of this relationship often being a lease arrangement.

Examples of Design for Long Life Service are:

- copiers in all formats (the toner system warns in time for renewal);
- power and heat stations;
- climate equipment in building environment;
- big infrastructure projects like tunnels, where all-in contracts for over 25 years are coming in practice (with design, construction, maintainenance and finance).

For organizations wishing to work in this way, the concepts of Integrated Design and Engineering must be implemented.

5.14 Design for Sustainability and Reuse

Design for Sustainability means two aspects to consider. Firstly are the materials that are to be utilised actually produced in a sustainable way (low energy costs, low CO_2, reusable). And secondly, is the product itself sustainable with low energy consumption, low CO_2 production, low pollution figures, etc.

Design for Reuse means asking during the design and engineering process what will be done at the end of the useful life of each part, assembly and module that is to be developed. This could be reusable through repairing, reusable through using in another place ('second life option'), reusable in terms of raw materials (e.g. modern cars are 98% reusable) 'Cradle to Cradle' is a relatively new vision concerned with the reusability of products and materials, the basic idea being that each product or material is raw material in a next phase of life. During the design process this idea has to be introduced into the process, so at the end of the life of each product it is known how it is be reusable.

Design for Sustainability and Reuse gives manufacturers the chance of creating *'delighters'* for their products, and the design team has to take notice for this important aspect.

5.15 Design for Production Costs and Life Cycle Costs

Design for Production Costs is manufacturing oriented and is of great importance for the future potential profit for the company itself. This is because the future profit of a company is tied up in the manufacturing costs of a design (product).

Design for Life Cycle Costs is also very important for future clients, in terms of their potential profit when they operate the product.

So we see that both costs categories are very important to consider. When we look to the production or manufacturing costs then the following cost elements are particularly important:

- material costs;
- labour costs;
- machine costs (incl. energy);
- logistic costs;
- coordination costs.

For each part, assembly, module and final product these elements are important, so in the design team a cost estimator will have an important role. For each part, etc. the manufacturing costs must be continually evaluated for decisions relating to issues such as robust design wishes, Life Cycle wishes, etc. The Hamburger model can be very helpful in developing a cost overview of the whole product that takes account of any conflicting wishes, see example in figure..... Studying solutions of competitors, patents, etc. can be very helpful in making the choices.

When we look at the Life Cycle Costs than generally the following elements are important:

- Investment costs (one time) with:
    - Purchase costs;
    - Installation costs;
    - Training costs;
    - Set of repair parts.
- Operational costs ('n' years of operation) with:
    - Labour costs;
    - Energy costs;
    - Maintenance costs;
    - Waste and environmental costs.
- End of life costs (one time) with:
    - Income from sold parts;
    - Costs of cleaning;
    - Costs of breaking down and removal.

When designing for Life Cycle Costs, the design team not only needs a manufacturing cost estimator but also an operational cost estimator. Also when designing for Life Cycle Costs, delighters can be brought into the design.

In the case of design for long life service, knowledge about Life Cycle Costs and long life finance has be the subject of special attention in the team because sometimes long life means terms of 25 to 30 years.

5.16 Design Process as Value Chain Business Model

Integrated Design and Engineering as a business process is only valuable if this process can create business, profit or in other words value. The place for this value creation process will be in the 'develop' plain (see figure 1.7 or figure 4.2); creating a value chain from raw materials to final product including the production chain. So, the end result has to be a product that fulfills the wishes of the client and also fulfillls the manufacturer's requirements to produce things effectively and profitably.

A value chain can be helpful in highlighting these two demands that require fulfillment. We will see that creating a value chain is completely in line with the LCE vision of IDE. Of course the process starts with ideas and lists the client wishes, then generates functions and products (architectures), whilst concurrent to this is the development of manufacturing and logistics in the chain, including the make or buy decisions. This means that when two or more product (architecture) options are in development, all the impacts for each product possibility on manufacturing and logistics are also developed. All these possibilities will be also calculated in terms of their manufacturing and Life Cycle costs. The winning concept is finalized with the complete picture.

This integrated development process is sometimes referred to as 'Lean product development', with innovation planning, function creation and product creation as the phases (see figure 5.17). During the innovation planning phase the main question is what are the (real) client wishes, and possible ideas for solutions are generated concurrently including the supply chain architecture, with the 'make' or 'buy' possibilities.

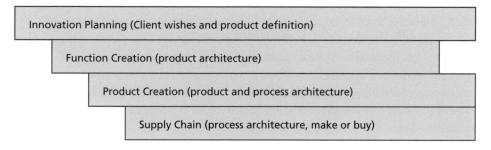

Figure 5.17 The phases of Lean product development

The function creation phase for developing the new parts of the best product solution will again include the supply chain requirements (manufacturing and logistics).

During the product creation phase the definitive value chain has been defined, including the product architecture.

5.17 Cost engineering

Cost engineering is a business model where the whole of the development process of a new product is driven from the initially defined performance indicators. E.g. the cost price should never rise above price X from the production line. Or the new product must generate a x% increase on the ROI. Or the cost price of the new product must decrease every half a year by 5%, etc. Cost engineering is often applied in very competitive markets like consumer electronics and the car industry.

The Hamburger model can be applied for cost engineering studies because for each function-product combination (each hamburger) a cost price can be generated. For each function-product combination there is usually more than one possible solution, so the optimum solutions can be calculated using value analysis techniques.

5.18 Design Kernel for Engineering

We will finish this chapter with a summary of all activities and methodologies that have to be, or can be, applied by the design team during the design and engineering phase of the Life Cycle Engineering process. We started this chapter by defining the sequential phases of this process. It is important to recognise the influence of manufacturing, installation, operating and maintenance factors or elements in the final design of a product. This is achieved through the use of a concurrent engineering methodology within the organization. Value chain thinking is very helpful in highlighting the scope and outcomes of the total development process for making a new product.

We will summarise below the phases of the Life Cycle for developing the **kernel** for engineering in order to give the methodologies we have introduced their place in the integrated design and engineering process.

Design and Engineering Kernel

Phase	Methodologies	Documents
Idea or Conception Phase	QFD (Cohen)	
	Brainstorming	
	Marketing research	
Result	Product definition with	
	one or more products	PDFDoc

Design Phase	QFD (Cohen)	
	EQFD (Clausing)	
	House of Quality	
	RAMSHE specifications	
Result	Product to be engineered	PDSDoc

Engineering Phase	Methodologies of Clausing	
	Pugh, Van den Kroonenberg, etc.	
	FMEA, FTA,	
	Robust Design Taguchi	
	Design for	
	Methodologies	
	Design for Sustainability	
	Cost models	
Result	Product ready to produce	PENDoc

Manufacturing Phase	Design for Assembly	
	Design for Costs	
	Test methodologies	
	(Prototype)	Test reports
Product	Product ready to sell	PMADoc

Installation and Commissioning Phase	Installation manual	PSPDoc
	Test methodologies (client)	Test reports
Result	Product ready to operate	Operation manual

Operational and Maintenance Phase	Ergonomics	
	Design for Long Life Service	
	Design for Maintenance	PSPDoc
Result	Long life product contact	Operation manual

End of Life or Out of Operating Phase	Design for Reuse	PSPDoc
Result	Out of service	

The documentation for Part Selection will comprise the following:

- PDSDoc for Product Design, containing all chosen solutions with their specifications (functionally and technically) and costs estimates (manufacturing and Life Cycle). In the case of a product family it is also possible to make a first draft product generator for the sales and sales engineering departments.
- PENDoc, for Engineering Parts, including drawings and containing, for all chosen solutions, the manufacturing and assembling instructions, all the technical details, materials, work methodologies and make or buy decisions.
- PMADoc, for Manufacturing and Assembling, containing all chosen solutions for manufacturing and logistics, make or buy decisions, manufacturing technologies, logistics.
- PSPDoc, for Product Support with Installation, Commissioning, Operating and Maintenance manuals, containing all the manuals for proper installation, tests for commissioning, starting up , operation, maintenance and after sales contacts (repair parts, guarantees, etc.) and end of life possibilities.

5.19   Product Configuration Model for a Product Family

In chapter 3 we have discussed the idea of generating a product configuration model on the basis of the Hamburger model. During this stage of the product development process a **product configuration model** can be made when we can consider selling a product family in place of one steady product (see figure 5.18). This means that the product family contains variations in things such as size, capacities and speed from the single product concept. For the sales department such a configuration model provides the possibility to respond very quickly to clients' requests, including the cost prices! Because all the standard solutions are already in the model, a quotation can be put together very rapidly (see figure 5.19, where an on line configurator for a belt conveyor is illustrated). Extras or special wishes will have to be paid for.

Manufacturers of belt conveyors can deliver conveyors with variations in width, in length, in the material of belts, in transport angle, weight of the product to transport, and in speed or capacity.

The following relational rules are applicable:

Capacity A (75%) can have: width a, length 1, 2, angle up to 30°, belts X, Y, Z, motor A.
Capacity B (100%) can have: width a and b, length 1, 2, 3, angle up to 30°, belts X, Y, Z, motor B.
Capacity C (125%) can have: width a and b, length 1, 2, 3, angle up to 30°, belts X, Y, Z, motor C.

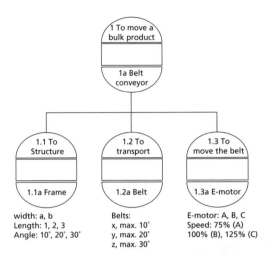

width: a, b
Length: 1, 2, 3
Angle: 10°, 20°, 30°

Belts:
x, max. 10°
y, max. 20°
z, max. 30°

E-motor: A, B, C
Speed: 75% (A)
100% (B), 125% (C)

Figure 5.18 Example of a Product Configuration Model

Note: For all these solutions in the family are the drawings etc. complete available, so that the production can start relative quickly.

Function /Solution	1.1a	1.2a	1.3a
1:a cap. A (75%)	width a		Motor A
cap. B (100%)	width a and b		Motor B (not A)
cap. C (125%)	width a and b		Motor C (not A, B)
1.a: cap. A	length 1 and 2		
cap. B	length 1, 2 and 3		
cap. C	length 1, 2 and 3		
1.a: cap. A	angle 10,20 30	X=10,Y=20, Z=30	
cap. B	angle 10,20 30	X=10,Y=20, Z=30	
cap. C	angle 10,20 30	X=10,Y=20, Z=30	

Table 5.4

So when a client wants to order a belt conveyor with capacity 75%, width a, length 2, angle 0°, the applied belt will be automatically X and so the motor will be A.

The order notation can be as follows: A-a-2-0-X-A

When the client changes its wishes in a later stage from angle 0° to 30° then the quotation automatically will changed to: A-a-2-30-Z-A.

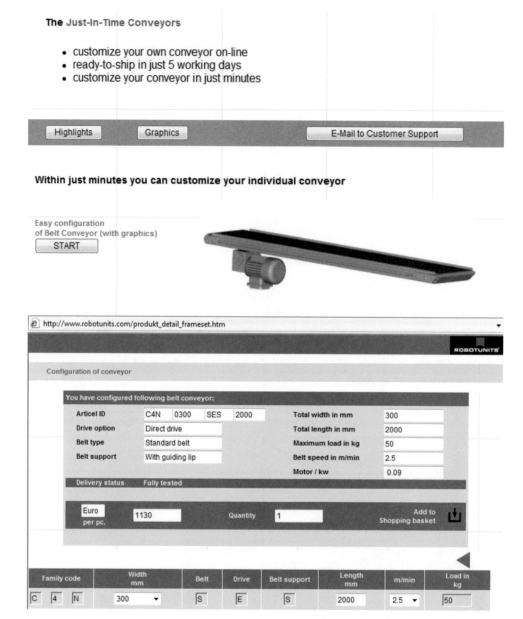

The Just-In-Time Conveyors

- customize your own conveyor on-line
- ready-to-ship in just 5 working days
- customize your conveyor in just minutes

| Highlights | Graphics | E-Mail to Customer Support |

Within just minutes you can customize your individual conveyor

Easy configuration
of Belt Conveyor (with graphics)
START

http://www.robotunits.com/produkt_detail_frameset.htm

ROBOTUNITS

Configuration of conveyor

You have configured following belt conveyor:

Articel ID	C4N	0300	SES	2000		Total width in mm	300
Drive option	Direct drive					Total length in mm	2000
Belt type	Standard belt					Maximum load in kg	50
Belt support	With guiding lip					Belt speed in m/min	2.5
						Motor / kw	0.09

Delivery status   Fully tested

| Euro per pc. | 1130 | Quantity | 1 | Add to Shopping basket |

Family code			Width mm	Belt	Drive	Belt support	Length mm	m/min	Load in kg
C	4	N	300	S	E	S	2000	2.5	50

Figure 5.19 Example of a configurator. You can assemble your individual belt conveyor and order it on line (http://www.robotunits.com)

Example

Life Cycle Costs on the level of part selection

The selection process for a bearing identifies two makes from a pre selection as potential candidates. Which of the two is the best? The cost price of A is €100 and the expected life time is 4 years; the cost price of B is €130 and the life time is 6 years. Each replacement of a bearing is €100.

The LCC of A (over 12 years) is: 3 × (100 + 100) = €600
The LCC of B (over 12 years) is: 2 × (130 + 100) = €460

When the installation contains 10 bearings that will mean the following difference:

Investment costs of: 10 × 30 = €300
LCC costs of: 10 × 140 = €1400 (over 12 years)

Example

Life Cycle Costs and sustainability

Life Cycle Costs calculations in the building environment can also provide sound answers about sustainability questions. Usually the integrated sustainable systems show higher investment costs but substantial lower energy consumption costs. Very often we found pay back periods of 7 to 10 years. We will calculate the LCC for two situations of a relatively small house with two different types of installations. Type A has a classic HR heat with additional ventilation and in summer a movable airco unit. The investment costs are €12500 and the energy consumption is about €2000 / year. For Type B these figures are €22500 and €1000 /year.

We calculate the LCC costs over 15 year (the life time of this type of equipment) and the pay back period:

	Type A	Type B
Investment:	€12500	€20000
Energy costs (15yrs)	€30000	€15000
End of life	pm	pm
Total	€42500	€35000

Pay Back Period B versus A = (€20000 − €12500) / (€2000 − €1000) = 7.5 years.

Summary

Guided by the Design and Engineering Kernel, we have described the process from the client wishes document PDFDoc to a complete and makeable product that fulfills the client wishes and also promises profits for the organization. During this chapter we have also introduced ideas from other well known writers on this subject. The product development team is encouraged to find their way by selecting one or more of these methodologies.

We have also seen the important role that a structured model like the Hamburger model can play during this stage of the development, especially in relation to the structured engineering database. Of particular significance is the possibility of building a product configurator on basis of the Hamburger model which can address products that are to be delivered as a product family.

Exercises

1. Make a Hamburger model, a *Fc*-FMEA and a FTA of a hydraulic system consisting of an oil container, pipes, an oil cooler, a pump, a switch and a hydraulic cylinder for axial movements.

2. The selection process for a power switch of a TL armature delivers two makes as potential candidates from a pre selection exercise. Which of the two is the best? The cost price of A is €5 and the expected life time is 3000 hours; the cost price of B is €16 and the life time is about 6000 hours. Each replacement of a switch costs €5. There is also a difference in power consumption between A and B of 0,01 KWh. The price of a KWh is €0,20.

3. For a part selection the requirements are:

    *   costs;
    *   material used;
    *   easy to transport
    *   easy to change.

    The wishes are:

    *   weight;   weight factor: 1
    *   well known relation;   3
    *   robust to operate;   2
    *   just in time delivery;   4

Four solutions came out of a first selection as possible candidates. An analysis using the selection matrix gave the following result:

Possible Alternative Solutions:	A	B	C	D
The requirements are:				
• costs	+	-	+	+
• material used	+	+	-	+
• easy to transport	+	+	+	+
• easy to change	+	-	+	+

The solutions A and D score both four times a plus, so they are the candidates for further selection.

The wishes are:
- weight: $1 \times (100)$ A, B, C, D are equal
- well known relation: $3 \times (100)$ A better than the others, rest equal
- robust to operate: $2 \times (100)$ C better than the others, rest equal
- just in time delivery $4 \times (100)$ D better than A, rest lower and equal

The results of the score of the wishes are:
- weight: $A = D = 50$ A, B, C, D are equal
- well known relation: $A = 70, D = 30$ A better than the others, rest equal
- robust to operate: $A = D = 50$ C better than the others, rest equal
- just in time delivery $D = 60, A = 40$ D better than A, rest lower and equal

a. Calculate the total score.

Total Score:		A	D
• weight:	$1 \times (100)$	.. × .. = ...	.. × .. = ...
• well known relation:	$3 \times (100)$	.. × .. = ...	.. × .. = ...
• robust to operate:	$2 \times (100)$	.. × .. = ...	.. × .. = ...
• just in time delivery	$4 \times (100)$	.. × .. = ..	.. × .. = ...

Total: = ...   = ...

b. Which part (A or D) is the best?

4. a   Make a Product Configuration Model for a pump system family with the following capabilities:

Frame sizes: I for A and B ; II for B and C
Pump sizes: A (50-90%), B (80-120%), C (100-160%)
Cost prices: A (100%), B (125%), C (140%)
E-motor: 1 (60-120%), 2 (100-160%), pump capacities
Cost prices: 1 (100%), 2 (120%), 3 (135%), 4 (145%)

b. Make a LCC calculation for two types of pumps (A and B) meeting the same
   functional specifications.

	Type A	Type B
Investment costs	€55000	€50000
Energy costs (4%resp 4.5% of investment /yr) × 15 yrs	€	€
Maintenance costs (3.5 resp. 4% of investment/yr) × 15 yrs	€	€
End of life (10%)	€	€
Total:	€	€

Literature

- Cohen, L., *Quality Function Deployment. How to Make QFD Work for you*,
  Addison-Wesley Longman Inc. 1995, ISBN 020163302.
- Clausing, D.,*Total Quality Development, A Step by Step Guide to World Class
  Concurrent Engineering*, ASME Press, 1998, ISBN 0791800695.
- Pugh, S.,*Total Design. Integrated Methods for Successful Product Engineering*,
  Addison-Wesley, 1991, ISBN 9780201416398.
- Ullman, D. G.,*The Mechanical Design Process*, McGraw Hill, 1997, ISBN 0071155767.
- Yang, K. and B. El-Haik, *Design for Six Sigma*, McGraw Hill, 2003,
  ISBN 0071412085.
- Clausing, D. and V. Fey, *Effective Innovation*, ASME Press, 2004, ISBN 0791802035.
- Kroonenburg H.H. van den. and FJ. Siers, *Methodisch Ontwerpen*, EPN, 1998,
  ISBN 9011045297.
- Beer, J.C.F. de. *Methodisch Ontwerpen*, Academic Service, 2006, ISBN 9039524556.
- Delhoofen, P., *Handboek Ontwerpen*, Wolters-Noordhoff, 2003,
  ISBN 9789001203375.
- Buijs. J. and & R.Valkenburg, *Integrale Product Ontwikkeling*, Lemma, 2005,
  ISBN 9789059313491.

Production, Operation, Maintenance and Continuous Improvement

6.1 Introduction

About 90% of all the development work of a design and engineering department is working on existing products. Guided by the ideas of clients, new market possibilities, new ideas or competition, existing products will be improved and redesigned. Reviewing and redesigning existing products to develop improvements in ease of operation and maintenance, energy consumption and sustainability, and the reuse of materials after end of life, will often generate more income than completely new products. Improvement methodologies like Six Sigma, Value Analysis and FMEA are applied in combination with the IDE philosophy.

Besides the development work on the product, the complete production process (from receiving order to delivery, including the logistic chains) also has to be the subject of continuous improvement. When we look to the business model of (see figure 4.2), we are acting now in the 'executing' plain with the main processes being: order engineering, product engineering, manufacturing and after sales activities. Remember this is the plain where the real income of a company is generated and is, therefore, of enormous importance to the continuity of a company. So, the main scope of this chapter will be: 'how to generate income through the continuous improvement of products and processes'.

We start by looking at the after sales activities of a product as source for generating continuous income over the whole life of a product for the company, and as input generator for ideas and all types of improvements from the client side.

Then we give a overview of possibilities for process improvement, e.g. by the lean manufacturing methodology

6.2 The After Sales Problems

'Help the client has a complaint' is very often the first reaction of a company upon the arrival of this complaint. In the worst case this compliant is ignored as a non problem, resulting in mainly unsatisfied clients as result. In modern marketing, clients complaints are not seen as a threat but as a challenge. They are a chance for achieving client satisfaction through product improvement. It is the challenge to make client complaints proactive and turn them into a tool for product development. This means that the receipt of complaints must be organized by a professional department and that this department not only has a role to play in following up these complaints to ensure client satisfaction, but also as an input translator of new client wishes.

6.2.1 The after sales as a business challenge

The existing range of products used by several clients can be an enormous source of information about the daily use and behavior of parts of these products, especially under different operating circumstances. For each manufacturer this information about the real life of the product provides a source for continuous improvement programs for these products. A special product support or service department that has daily contact with clients is normally the helpdesk organization for clients and is the logical place for the input and the processing of this information. This is in addition to the direct help and support for actual problems.

Again this information has to be stored in a structured way and once again the Hamburger model can be applied to store all the data about problems. With the help of this model it is possible to show problematic parts, assemblies and modules on all levels of the functional decomposition tree. By giving different colors to the functions on all levels to reflect the situation of the function-solution, a good overall view can be generated about the satisfaction levels of the clients.

The quality figures of the functions in the tree can be colored as follows (see figure 6.1):

- Red for: Troubles (the problem has not been resolved), direct repair action and direct action for improvement.
- Orange for: Can be better, room for improvement.
- Yellow for: Normal, works as expected.
- Green for: No problem, works even better than expected.

Information about part behavior can be continuously used for improvement actions of the part (assembly, module) design by the design and engineering department.

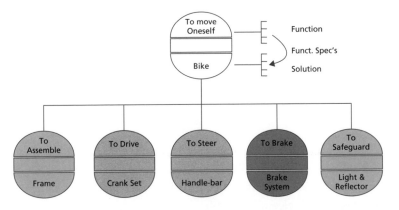

Figure 6.1 Hamburger model for an overview of possible problems

In addition to information about good and bad operating behavior of the product, which is mostly dependent on the operational context, the manufacturer has to realize that a proactive after sales or product support department can generate a lot of income over the Life Cycle of a product by organizing a 'whole life' client contact.

Potential areas for after sales include:

- services by distance for condition monitored parts, including the part logistics;
- control by distance (information connection via internet) for installation behavior;
- complete integral maintenance packages;
- selling improved parts (e.g. less wear or better performance);
- development of modification kits for better performance of the product (e.g. higher output);
- organizing client contacts to identify new ideas for enhanced performance and new functions to be added;
- organizing a client community for the exchange of experiences and new ideas;
- *second* and *third* life of a product.

6.2.2 The after sales as a business

The starting point for an after sales business activity is the contract and order engineering step of the selling process of a product. At that point the client decides what type of contract they want. Possible contracts are:

- only selling a product, no after sales;
- selling including installation and commissioning, delivering parts on demand;
- selling including a service contract for an agreed period of years;
- selling including a full service and operating contract for an agreed period of years.

Depending on the contract, a product will be designated as price oriented for selling purely on price, and service oriented for selling the product together with after sales services. So the design processes for these types of product service combinations have different implications for the design team. This is especially the case in assembly to order, engineering to order and single luxury product manufacturing where a long life after sales business is possible.

In the case of mass consumer manufacturing after sales is mostly replacing at low (no) cost any faulty products and also delivering consumer products for these products, such as toner and print paper for PC printers, coffee pads for coffee machines, shaver creams and after shave products for electric shavers.

In all cases after sales business is very profitable because the prices for these services are more or less exclusive and are usually less sensitive to lowest cost price erosion effects.

Case 1: shipbuilding yard for dredgers and dredging equipment
A shipbuilding yard for dredging equipment sold this equipment inclusive of all the drawings. They considered the new building activity as the main activity and the after sales service was very poorly organized. This situation continued until they suddenly discovered that in places where a lot of dredging activities had taken place, local machine shops made their equipment by imitating their designs (with the original drawings!) and selling these at prices which were double the level of the new equipment prices. As a result, a lot of business was missed during a lot of years.

Case 2: diesel engine factory
A diesel engine factory that manufactured industrial diesel engines stopped their production 25 years ago, but still make and deliver a service organization providing original parts for these engines at high profits.

Case 3: coffee pads
The companies Philips and DE deliver a special coffee machine (Senseo) which applies coffee pads. The profit on these pads gives Philips extra income and DE a nice margin. So both companies enjoy a profitable situation.

Case 4: printer toner
Perhaps the best known after sales application is the toner for printers. The cost price of toner material is very low, but by designing special toner containers customers have to buy this special package for each printer. For manufacturers the margin on this toner package is very high. Typically they lose money on selling the printers and make high profits on the toner.

6.2.3 The after sales engineering

During the manufacturing phase of a product, which can cover a number of years, client complaints and new client wishes or ideas are all continuously passed through to the engineering department. For this department (and of course for the organization) it is of enormous importance to follow up these new opportunities to improve the product. Again the Hamburger model can be applied to visualize the new wishes, specifications, functions, etc. The engineering department will establish a sub department especially for collecting these complaints, wishes and ideas. It is very useful in terms of client satisfaction if client contacts can be proactively organized with the help of service letters or instructions, and even with specially developed service modification kits for improvements to existing products that have already been delivered.

Service letters inform clients about some of the more generic complaints and what the manufacturer has done to solve these. Service instructions can clients help to improve the specifications of the existing products. Service modification kits are specially developed for existing products to improve the specifications to the new standard.

After sales engineering can also be termed as *'design for continuous improvement'* and will always, when organized in a proper way, not only be very useful for contacts with clients, but is also prove a very profitable activity with relatively high margins. This client input can be also very helpful for formulating the specifications through further development of both existing and new products.

6.2.4 How to organize after sales engineering?

We have noted that *after sales engineering* can be a source of continuous income and profit for a company, so the question of how best to organize this activity is of importance. An effective solution is the formation of a clients' complaints team, responsible for collecting the incoming complaints and ideas, for prioritizing the client help actions, for initiating improvement activities and for proactively providing client information about the progress of the treatment of complaints. The place of this team is in the proximity of the after sales department which is in daily contact with clients.

The composition of this after sales engineering organization will be made up of experienced people from service and construction departments, who will have experience of daily client contacts. The most important tasks of this department will be:

- collecting all client complaints and ideas:
- initiating immediate action on urgent complaints and communicating daily with the client about progress of possible solutions for eliminating the complaint;
- producing statistics from complaint data and presenting this in the form of a Pareto diagram;
- prioritizing the complaints from the Pareto outcomes;
- formulating improvement projects for eliminating complaints and for developing new ideas;
- delivering staff members and/or project managers to execute these improvement activities;
- carrying out the client information activities via service letters and service instructions;
- selling all possible long life services, including modified parts in products.

6.3   Improvement Activities

A lot of improvements and improvement activities are possible. It is very important for each company to pay attention to these activities. After prioritizing the improvement activities we will look at the following:

- activities for eliminating complaints related to existing products;
- activities for designing improvements to products (originating from client ideas or market developments);
- activities to improve the existing production line (lower costs, higher output, better quality, etc.).

6.3.1 The activities for eliminating complaints

Activities for eliminating complaints can be divided in two main categories:

- Eliminating a complaint for one client (often from a special operating context). This complaint can be solved usually by design activities in the repair sphere that provide solutions for a specific situation at a single client.
- Eliminating a complaint that has been received from a number of clients over a certain period of time and which often means a serious problem with one of the functions of the product. In other words the product does not properly fulfilll a function specification and comes in a state of functional failure.

These activities generally provide higher client satisfaction and some potential business opportunities.

6.3.2 The activities for designing improvements

Activities for designing improvements can also be divided in two categories and can come from market or client ideas:

- Improvements in existing products (functions), e.g. higher production rates, better part performance, lower cost price, etc;
- Adjust new function(s) to existing products, e.g. control on distance, condition monitoring techniques.

These activities will typically generate new income and reach new clients.

6.3.3 The activities for improving production processes

Activities for improving production processes have also to be a continuous improvement activity driven by the market and the clients. Lower cost prices (market), higher quality (clients) and higher production output are drivers for improvements.

The Lean philosophy for process improvement is here also applicable.

Note: Organizing improvement activities is a very important business activity for each company in order to generate as much money as possible from the existing products (families) over the whole Life Cycle for the continuity of the organization. A nice extra benefit of these activities is that they also provide high client satisfaction and therefore form a good basis for further business development.

6.4 Analysis Methodologies for Eliminating Complaints

In chapter 5 we introduced a lot of methodologies for product development, we mentioned: QFD and EQFD, the methodologies of Pugh and Van den Kroonenberg, Robust Design of Taguchi, TRIZ, FMEA, FTA as the most important ones. For the after sales engineering we will introduce some other very helpful methodologies like: Brainstorming, Pareto analysis, the Five Times Why Question method, the Keppner & Tragoe method, the Fishbone diagram and Root Cause Failure Analysis (RCFA). For each given situation the design team has to decide which methodologies they will apply.

6.4.1 Brainstorming

Brainstorming is a technique to investigate a problem and to find solutions in a creative way. The methodology is well known and only mentioned here as a useful method to generate ideas. The outcome is mostly a list of possibilities of solutions to problems e.g. failures or client complaints. These ideas have to be subject to further investigation by eliminating possible solutions in order to find the real solution (of a problem).

6.4.2 Pareto analysis

The **Pareto analysis** is very helpful in prioritizing the incoming complaints from clients. The method starts with collecting data about complaints. The different types of complaints become names and numbers and over a certain period of time the complaints are collected. All the data is ranked according to the amount of complaints for each number. Bringing this data into a diagram we can quickly see which complaints have the highest amount of failures and which is number two, number three, etc. Very often 20% of the numbers of complaints will account for 80% of the troubles in time or money. So in the first line of complaint analysis we can concentrate on these 20% for prioritizing the engineering work.

Example

An analysis of a total of 100 complaints with ten different complaint (failure) forms gave the following outcome:

Place 1:	Complaint 5	57 failures = 57%	57%	xxxxxxxxxxx
Place 2:	Complaint 8	23 failures = 23%	80%	xxxx
Place 3:	Complaint 1	9 failures = 9%	89%	xx
Place 4:	Complaint 4	4 failures = 4%	93%	x
Place 5:	Complaint 9	2 failures = 2%	95%	–
Place 6:	Complaint 2	1 failures = 1%	96%	–
Place 7:	Complaint 6	1 failures = 1%	97%	–
Place 8:	Complaint 10	1 failures = 1%	98%	–
Place 9:	Complaint 3	1 failures = 1%	99%	–
Place 10:	Complaint 7	1 failures = 1%	100%	–
			Pareto results	
Total:		100 failures = 100%		

So the complaints 5 and 8 account for a total of 80 failures (80%), which is not an unusual outcome.

It is possible that in order to solve a complaint we have to decompose the functional structure around the complaint and then find that a new Pareto analysis has to be made on a deeper level of this complaint, or even still deeper on a third level. The result will be the parts which give the most problems and these parts are then subject to improvement activities.

6.4.3 Methodology of Kepner and Tregoe

The **methodology of Kepner and Tregoe** finds its origin in the fact that they found when analyzing reactions of teams of operators in a power plant on similar failure situation, that under pressure some teams solve the problem of a failure in a reasonable time, some teams take longer and some teams actually make the failure situation even worse! They discovered also that the way in which successful managers solve daily problems under pressure is the same as successful teams did. So they developed a methodology around solving problems under pressure when a deviation of a given norm or standard occurs. A lot of client complaints falls into this category.

Kepner and Tregoe have discovered that there are four basis routines for solving problems under pressure when deviations of outcomes to a norm or standard occur:

- The first basic routine is the question: hey, what is happening now?, or: *situation determination*.
- The second question is: why does this happen?, or *problem analysis*.
- The third question is: what has to be done?, or *decision analysis*.
- The fourth question is: how (can we solve) and what-if (happens) questions, or *potential problem analysis*.

Good managers and teams are very strong in problem analysis, where they can place a problem in the right case.

General approach methodology
Step 1: Formulate what is the problem.
When a problem which a deviation to a standard comes on the table then firstly the situation should be determined. It is important to formulate the problem and commit it to paper before handling it.

Step 2: Specify the problem
Now we try to specify the problem deviations in two ways, first the 'Ist' and then the 'Ist Not'.

Analyzing the 'Ist' situation
By going through a series of four questions we further investigate the problem.

The first question will always be to identify the problem: **what** asset has a deviation, or what is the deviation? By writing down as soon as possible after finding the problem, the formulation and specification of this problem, this 'what' question is usually already answered! We can go further or deeper to formulate specifications around the problem.

The second question is: **where** (in which place) is the deviation found? This question will specify the exact place or area where the deviation is found (and where it is not).

The third question is: **when** is the deviation found? The 'when' question provides information about the time that the problem or deviation was found. When was the first time and when was it subsequently found, or when in the history of an asset was the deviation first noticed.

The fourth question is: **how big** is the problem, how many assets have this deviation? Answering this question will generate information about the size of the problem. Identifying how many assets out of the total indicates the scale of the problem, or how many deviations are found on each asset.

Analyzing the 'Ist Not' situation
After analyzing the ´Ist´ situation we will apply the same questions in the 'not could happen' form.

The first question will now be: what are the (similar) assets with no problems but which could be have problems also, or what could be the deviation?

The second question is now: where has the deviation taken place or where could be the place of the deviation?

The third question: when had the deviation taken place for the first time?

The fourth question: how many assets would have this deviation?

In table 6.1 we will summarize the questions of the ´Ist´ an ´Ist Not´ situations.

Problem Specification	
Ist	Ist Not
What is the asset with a deviation? What is the deviation?	**What** is the asset which could have troubles but which has not happened until now? What could be the deviation?
Where was the asset when the deviation was noticed? Where is the deviation on the asset?	**Where** could the deviation also have taken place? Where could be the place of the deviation on the asset?
When is the deviation found for the first time? **When** is the deviation found after the first time? When in history of the asset was the deviation found for the first time?	**When** had the deviation also taken place for the first time? When had the deviation also taken place after the first time? When is history of the asset could the deviation also have taken place for the first time?
How many assets have this deviation? Which part of the asset has the deviation? How many deviations are found on each asset?	**How many** assets could have this deviation? Which part of the asset could have the deviation? How many deviations could be found on each asset?

Table 6.1 Summary of ´Ist´ an ´Ist Not´ situations.

Step 3: Check possible causes with the problem specification.

During this we check possible causes from the outcomes of the questions in the problem specification. At first we try to formulate possible causes by:

- using our personal experience to formulate one or more causes;
- by looking at differences (one machine has troubles and other ones have not);
- by looking at sharp contrasts (sudden changes in operational figures);
- by looking at changes during the operating time (other material, other temperature);
- by possible cause of the working method itself.

For each possible cause obtains that it has to fulfilll a yes to all the questions or in other words: ´how can this cause produce this result´ or can this (possible) cause explain indeed **What** happens, **Where** it happens, **When** it happens and **How much** it happens, and also **What not** happens, **Where** it **not** happens, **When** it **not** happens and **How much** of all **not** happens.

In other words: ´can a possible cause explain everything?´ If our answer is yes then there is a good chance that this cause is the real problem and we can then usually solve the problem. It is also possible that we find two causes that both explain everything, in which case a further investigation into the specifications has to take place.

Step 4: Investigate possible differences
4a Differences in outcomes of operation of similar systems.
This step is often very useful when some similar machines are operating in the same condition. Investigating the differences between the machines which are working properly and the machines with troubles can often generate the possible solution to the problem.

4b Sharp differences
When the behavior of a system or machine changes very quickly then the main issue will be to look at the possibilities of change, what is happening before and after the change.

4c Possible causes from change in methodology or differences in methodology
Also changes in the methodology of working can introduce unexpected failures.

Note: in steps 4a, b and c, we are specifying the possibilities as mentioned in step 3.

Step 5: Verification
Before starting the repair activity the investigators of the problem have to verify the outcome, so the solution that has been discovered has to be proven by checking again all the established information in relation to this solution.

Example

Chocolate storage tanks
Step 1: Formulate: what is the problem?
In a chocolate factory 6 new chocolate storage (mixing) tanks are installed in the newly built hall. The temperature of the chocolate in the tanks has to be at least 50 °C and can vary between 50 and 60 °C. The tanks are heated by a hot water system controlled by a 3-way valve with an entry temperature of 75 °C and an outlet temperature of about 65 °C. After the first run of starting up the heating system of these tanks it appeared that in one of the tanks the temperature of the chocolate rose very slowly and that the required temperature of 50 °C was not reached.

What is happening here?
The problem is that in one of the six tanks the temperature of the chocolate does not become higher than 40 0C (instead of the required 50-60 °C).

Step 2: Specify the problem
We will specify the problem by applying the Problem Specification table (see table 6.2)

Problem Specification	
Ist	Ist Not
What is the asset with a deviation? Tank 2. What is the deviation? Temperature chocolate under the required 50 °C.	What is the asset which could have troubles but which has not had them until now? Tanks 1, 3, 4, 5 and 6. What could also be the deviation?
Where is the asset when the deviation was noticed? The new hall, a brand new mixing tank. Where is the deviation on the asset? Chocolate temp. too low (below 40 °C) Water inlet is done by 3-way valve and control-led by chocolate temperature controller.	Where could the deviation also take place? Not applicable (other tanks). Where could the location of the deviation be on the asset? Not applicable.
When is the deviation found for the first time? After the first starting up. When is the deviation found after the first time? Not applicable. When in history of the asset is the deviation found for the first time? After the first starting up.	When could the deviation also have taken place for the first time? When had the deviation also taken place after the first time? When is history of the asset could the deviation also have taken place for the first time?
How many assets has this deviation affected? One tank, tank number 2. Which part of the asset has the deviation? Outlet temperature controller, chocolate temp. How many deviations are found on each asset? One.	How many assets could have this deviation? All six tanks, but the others are working properly. Which part of the asset could have the deviation? How many deviations could be found on each asset? Not applicable.

Table 6.2 Problem Specification

Step 3: Possible Causes
A small team of specialists was formed to tackle this problem. The team members decided that two possible causes could be responsible:
- water flow is too low;
- control system is not working properly (temperature gauge or temperature controller).

Step 4: Investigate possible differences
4a differences between systems
Because all the other tanks work properly within the design limits the system of tank 2 must have some differences in working.

The team decide to control in parallel the temperature gauge of the control system and the water flow though the tank; the result of this test is that it is functioning properly, no mismatch of temperatures.

By dismantling the 3-way valve (to control the flow) the mechanics found the rests (remains) of the plastic safety prop, which had the effect of practically blocking the flow. So the cause of the problem was located.

Step 5: Verification
By mounting the 3-way valve in a proper way the installation was started again and was functioning now according to specifications without any trouble.

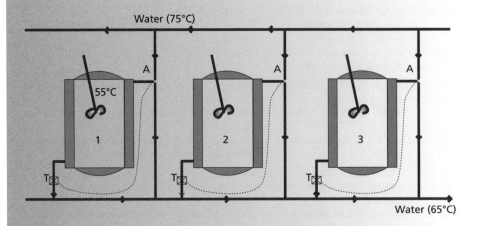

Figure 6.2 Mixing tanks (schematic)

Figure 6.3 Example of mixing tanks for the production of chocolate (© Delver S.p.A., Modena, Italy)

6.4.4 The Five Times 'Why Question' analysis

A methodology more or less similar to the Kepner and Tregoe methodology to find a cause for a problem or deviation is the **Five Times Why Question analysis**. By asking and answering the question 'why' five times, the deeper lying causes of a problem can be found. Through checking these causes against possible causes, the real cause can be found. The methodology is as follows:

Step 1: Describe the problem and the problem specifications.
Step 2: Ask 'why' to formulate a possible cause.
Step 3: Register the possible cause(s) in the table.
> Step 4: Check the possible cause(s) to the problem specifications or do a test.
> When the outcome of the check or test is True, give the cause a **T** in the register and go to step 5.
> When the outcome is not True give the cause a **X** and go back to step 2 (another possible cause).
Step 5: Repeat the steps 1 to 4 five times.

Example
Car did not start

W1	T1	W2	T2	W3	T3	W4	T4	W5	T5
Accumulator empty	X								
Start motor	T	Short-circuit	T	Water	T	Isolation	T	Replace	T
		Cable not fixed	X						

Table 6.3 Register of the Five Why questions

The isolation of the wires of the start motor were altered and caused a short-circuit, so the replacement of this motor was the only solution.

6.4.5 Fishbone or Ishikawa diagram

The **fishbone diagram** or cause-effect diagram (see figure 6.4) can also be used to solve problems or deviations. Using the brainstorm methodology all possible causes are analyzed. It can be used in combination with the methodology of Kepner and Tregoe and the Five Why questions.

The Method
Step 1: Define (describe) the problem.
Step 2: Organize a brainstorm session about all the possible causes of the problem.
Step 3: Develop a scheme with the four M´s (Man, Material, Machine and Method).
Step 4: Place the subjects of the brainstorm in the scheme and try to find the deeper causes using the Five Why questions method.
Step 5: Look for relationships and mark or select the causes which are probably the most important.
Step 6: Investigate the selected causes.

Note: Check the logic by reading back and asking: 'why does that problem result in that cause, or why does that cause result in that problem?'

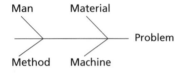

Figure 6.4 The fishbone or cause-effect diagram

6.4.6 Root Cause Failure Analysis (RCFA)

Principle
RCFA is a problem solving methodology intended to analyze the causes of specific (hidden) failures and to arrive at possible solutions for eliminating these failures. The process of RCFA is focussed on investigation, analysis, rectification and elimination of (system) failures. The methodology can be used to find incidents of hidden failures and potentially dangerous situations. The strength of the methodology is to identify the hidden causes, which are responsible for the failures (breakdowns).

Methodology
The following steps show the process of the RCFA methodology:

- Priority step
- The first step involves the selection of the priority of the failure to be solved. 80% of systems failures are initiated by 20% of the possible failure modes (Pareto analysis). A serious failure problem selected for the RCFA process typically relates to environmental, safety, production or quality issues.
- Analysis
- In this step the failure will be analyzed by means of a *´failure tree´* to get a realistic picture of the source(s) of the failure modes (FTA is Fail Tree Analysis).

The failure tree (a logical set up of cause and effects) is set up as follows:

Failure Event	Failure Mode A	Hypothesis A	failure mode A1, etc
			failure mode A2, etc
	Failure Mode B	Hypothesis B	failure mode B1, etc
	Failure Code C	Hypothesis C	no further steps

A small team undertakes deeper analysis (see also the functional decomposition models) in the system by asking which failure modes are possible, what are the reasons for these modes (hypothesis) and what are the failure modes (A1, B1, etc.) behind the first failure mode A. The hypothesis and formulation of the failure modes is then repeated for the third layer A11, B11, etc. The process continues until the real failure mode that solves the main failure is found.

- Recommendation
  The team formulates solutions to eliminate the solved failure event. It can also give recommendations for further eliminations of other failure events (priority).

Remarks

1 The execution of this type of analysis has to be done by a small group of people who are directly involved (operators and mechanics) assisted by experts.
2 The RCFA analysis methodology is an example of a methodology that can be applied to eliminate failures.

We have seen that there a lot of possibilities to tackle client problems arising out of deviations or failures. The result will usually be a solution to this problem and sometimes more than one solution. For a manufacturer the question to consider is whether the problem is a specific one off problem or if more clients are likely to be confronted by it. In the first case the client can be satisfied by solving their problem, whilst in the second case the manufacturer will have to communicate with all the clients, informing them of the possibility of this problem affecting them and providing details of the available solution. In the second case there is the opportunity to generate extra business by offering new solutions in the form of a repair modification kit.

6.5   Business Model for Improvements

We have seen that as part of handling complaints, proactive client contact can also bring
a lot of ideas and sometimes even solutions for improvement in product specifications,
or even new extra functions. The final point can be in the area of delighters. When we
look at the Hamburger model, the ideas can be placed in the model as follows:

- New or improved specifications in the specification list, e.g. proposal to improve the
  output.
- New solution of a function means an extra solution of a function, so the under side
  of the Hamburger includes an extra solution piece.
- New extra function means a complete new Hamburger somewhere in the scheme.

For all these types of proposals a quick scan has to be carried out and a decision to
develop this proposal has to be made and than prioritized. The client responsible for
generating the idea must be informed of the outcome as quickly as possible. Usually
they are the first client to buy this newly developed product.

Each proposal has to be checked about:

- market potential;
- development time;
- investments;
- cost price and selling price;
- Life Cycle Costs.

After this check for each proposal or idea, a plan of work for the further development of
these ideas has to be made.

6.6   After Sales Engineering Process

The after sales engineering process is, in general terms, the same as the engineering
process of a new product, although the starting point is usually an existing product. The
documentation is available for this product and the data from the product documents is
the starting point for the engineering process, in combination with a client complaint
or idea. Central to the start of the process is the functional decomposition of the
existing product, the Hamburger model and an investigation of the main function,
the specifications and the sub function(s) which relate to the complaint or idea for
improvement. The functional analysis can give direction to possible solutions for
solving the complaint or exploiting the ideas.

Depending on the situation, a full engineering process (see chapter 3) or a streamlined one can be followed.

We always start by applying the documents PDFDoc and PENDoc and investigating the content with regard to decisions made about the choice of functions, solutions and specifications. Then we have to decide whether to go for the full development scheme (described in chapter 5, see § 5.18 Design Kernel for Engineering), or to do the development process in a quicker way by going straight to possible solutions and starting directly with the part engineering. In all cases we have to investigate the patents involved and what solutions our competitors already have.

The result the process will be always:

- a revised PDFDoc;
- a revised PENDoc;
- revised PMA- and PSPDocs.

Example

Metro

A Pareto analysis was executed over the first period of running of a new metro (underground, subway) (see figure 6.5). The outcomes identified two main failures, the climate installation and the speed sensor. An improvement team (a small group of three people) was formed to investigate the problem of the speed sensor. They put together a decomposition model (Hamburger model, see figure 6.6) and a Fc-FMEA. By applying the critically matrix (see figure 5.7) they found that the cable of this sensor was not mounted properly and caused broken cables at uncertain times (see red Hamburger in figure 6.6). They proposed a simple modification by fixing the cable better and also mounting a special connector. After this simple modification, with a pay back period of 1.4 years, the failure disappeared.

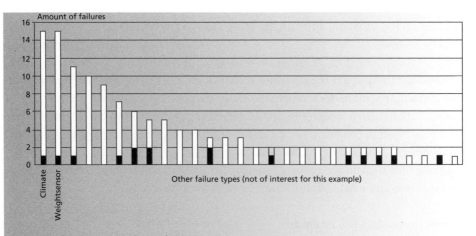

Figure 6.5 Pareto analysis of metro train

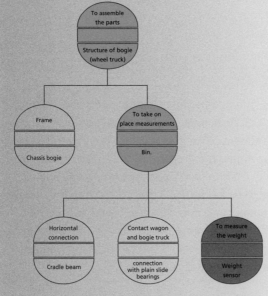

Figure 6.6 Red Hamburger for trouble shouting

Summary

The exploitation of after sales activity to generate solid business income is still an unknown field of operation for many companies. In this chapter we have introduced the after sales activity not only as a source of income but also as a very valuable source of ideas for the improvement of existing products. In particular, improvements to eliminate failures and complaints, and to increase the functionality of products.

Exercises

1. Apply the ´5 times Why Question´ methodology to the example of the chocolate storage (mixing) tanks.
2. Develop a business proposal for a new extra function for a pump, e.g. an electronic speed controller (for saving about 2% energy or €300/year).
3. A Pareto analysis was executed over the first period of running of a new metro train (see figure 6.5). The outcomes showed two main failures, the climate installation and the speed sensor. An improvement team was formed to investigate the problem of the climate installation. What type of improvements could they find?

Literature

- Yang, K. and B. El-Haik, *Design for Six Sigma*, McGraw-Hill, 2003, ISBN 0071412085.
- Kepner, C.H. and B.B. Tregoe, *The Rational Manager*, McGraw-Hill, ISBN 0703411753.
- Rummler, G.A. and A.L. Brache, *Improving Performance*, Jossey-Bass Publishers, 1995, ISBN 0787900907.

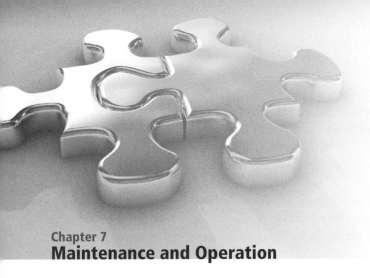

Chapter 7
Maintenance and Operation

7.1   Introduction

A product which is working in its operating context will generate the need for restoration or repair activities to maintain the functional specifications of this product. So the design of a product and, through this design, the approach required for repair (or maintenance) activities has to be optimized in order to achieve optimal results in its future operation. We will look at a methodology for checking existing and new products or installations on their maintenance behavior and also to show the possibilities for improvements in this field.

Two main phenomena or processes characterize this need for repair activities, the maintenance degradation process and the maintenance failure process. The outcome of these processes is an amount of hours, the total of all repair activities, what we can term as the ´Maintenance Need´ of a product. The main goal is to optimize or to minimize this amount of hours. Each process has its own maintenance strategies for optimizing the maintenance process.

In general, when operating an asset, this asset will generate the need for repair stops to execute repair work because of the wear of parts and degradation of its functions. We indicate this process of wear and degradation as the maintenance degradation process. In addition, unplanned failures will generate the need for repair activities, and this is called maintenance failure process. Crucial to selecting the right strategy is the degradation behavior of the asset, or a part of this asset, during the time of its use or its operation, in other words the wear patterns. If these patterns are unknown then a correct choice of strategy is impossible.

The degradation process of operating an asset will mean that preventive repair strategies like Condition-Based Maintenance (CBM), Time-Based Maintenance (TBM) and Failure-Based Maintenance (FBM) are possible as maintenance repair strategies. However, the failure process will generate unplanned and undesirable break downs of the asset and this process will always generate repair activities that are not preventable. It requires a special strategy, usually provided by the failure elimination modification strategy (FEM). The repair activity after a. unplanned failure break down is indicated as ´repair after failure break down' (RFB). For both processes we give an overview of their influence on the availability (A) of an asset, an important factor for asset efficiency.

After the introduction of maintenance as a process, we introduce the most common preventive maintenance repair strategies and the conditions and circumstances in which to select them. Their outcomes are the basis for a real and structured maintenance plan.

The product development team (especially the representative of the product support department) has to realize which parts are sensitive to wear patterns, how these parts can be easily exchanged and which parts can be subject to failure. The right Fc-FMEA can provide a lot of answers for these questions.

7.2   Maintenance as a Process

Maintenance is an activity, or a series of activities, necessary to bring an asset back to a desired level of operational demands. When an asset is placed under certain levels of operational demands it may be unable to fulfill the function(s) for which it was originally selected. The use and/or operation of an asset will generate processes of wear and tear or failure(s), and so the requirement will arise for restoration activities which will bring it back to the desired level of operational demands. This is an ongoing process or movement: it will start at a certain level of operation, with the asset having to stop at a certain level of output through wear, or a failure requiring a total stop or break down of the asset. For both reasons, restoration or repair activities will need to take place.

In other words through use and operation, assets will always generate a **maintenance need** for:

- repair hours of the assets and this will result in a certain availability of the asset;
- repair man hours for craftsman (mechanics, etc.);
- spare parts used by these repair activities;

Maintenance consists of two main fields or processes for generating restoration work:

- restoration work generated by wear or by degradation processes;
- restoration work generated by unplanned failures (failure processes).

For both types of maintenance restoration or repair work we will look at the specific behavior of the process and also at the possibilities for restoring the asset.

7.2.1 Maintenance required through degradation process

When using or operating equipment or assets parts of equipment such as bearings, V-belts, etc. there will be a loss at certain times or certain levels of performance due to wear phenomena. This means that at regular times these parts have to be exchanged and replaced by new ones. The **degradation process** is a constant movement between a new situation and, after some time, a worn out situation and it takes the form of a meandering process of replacement-wear-replacement (see figure 7.1).

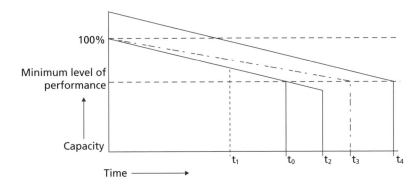

Figure 7.1 The degradation process

When we look in more detail at this figure, some remarks can be made. First the level of performance at which time (t_o) the change of part takes place is often not well known and can be part of improvement studies. This can result into two situations, whether the time t_o is too short and can be enlarged to t_2 or if the time t_o is too long and has to be shorted to t_1. In the case of too short, the improvement comes from the longer time t_2 which will result in less stops and a longer operational working time of the equipment. In other words, in a higher level of availability. In the case of too long the improvement comes from the shorter time t_1 which can result in a higher quality level at the moment of stop at t_1. Both situations, higher availability and higher quality, will result in a higher production output or a higher amount of quality products, or in other words in a higher level of income (or turnover) which results directly in higher sales and usually in a higher profit for an organization. In general, repair activities resulting from the degradation process can be planned and are be described as **Planned Preventive Maintenance (PPM)**.

The degradation process of repair-wear-stop-repair-etc- is illustrated in figure 7.2.The mean time *between* repairs is indicated by MTBR and the mean time *to* repair by $MTTR_R$. These two ´mean times´ give the following expressions:

For the availability due to planned maintenance:
$A_R = MTBR / (MTBR + MTTR_R)$ in %

For the number of stops for planned maintenance:
$\lambda_R = TOT / (MTBR + MTTR_R)$ in numbers per unit time (mostly year)

For the total time to stop (TDT = Total Dead Time) for planned repairs:
$TDT_R = \lambda_R \times MTTR_R$ in hours, is also a measure for availability
see:
TOT (= Total Operational Time = max. 8760 hours/year)
TPT_R (= Total Production Time by repair) $= TOT - TDT_R$

To improve the effectiveness of the maintenance activities the following actions have to be taken. To achieve a higher A_R, than lower λ_R and lower TDT_R, or higher the MTBR and lower the $MTTR_R$.

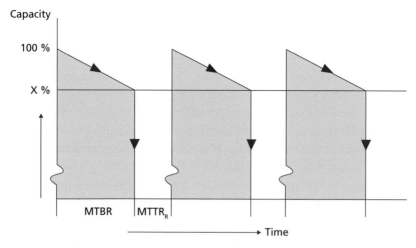

Figure 7.2 Degradation process of repair-wear-stop-repair, etc.

Example

In analyzing a production system the following data was found:

MTBR =1035 hours and $MTTR_R$ =60 hours.

1 Calculate A_R, λ_R, TDT_R, TPT_R (with TOT = 8760 hours)

A_R = 1035 / (1035 + 60) = 0.945 = 94,5%

λ_R = 8760 / 1095 = 8 times/year
TDT_R = 8 × 60 = 480 hours/year

TPT_R = TOT – TDT_R = 8760 – 480 = 8280 hours/year

Exercise
Calculate this figures as MTBR = 2110 hours and $MTTR_R$ = 80 hours.
Answer: A_R = 96.3%, λ_R= 4, TDT_R= 320 hours, TPT_R = 8420 hours.

7.3   Maintenance required through process failure

When operating an asset another phenomena like degradation or wear and tear can take place, namely failures can happen at uncertain times. In this case a failure means a total break down of the asset, so that this asset is unable to fulfill its function. In practice this usually means a break down of the production function (in terms of quantity and/or quality). After the break down a restoration activity has to be carried out, to bring the asset back into a situation where it is able to fulfill its function again. Central to the maintenance failure process is the **reliability (R)** rules relating to the amount of failures in a certain time. The most commonly occurring reliability behavior for systems (about 95% of all cases) is ruled by the expression: $R(t) = e^{\lambda t}$, in which $\lambda(t)$ represent the amount of failures in a certain period of time (see figure 7.3). In this case the λ (t) is constant over a certain period of time ('constant failure speed').

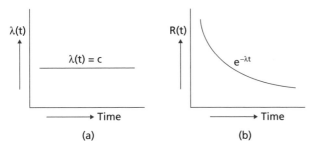

Figure 7.3 Reliability behavior for systems

By decreasing the λ in a certain period of time, the Reliability R will increase. So the general policy will be to decrease the amount of failures over a certain period of time. At a point in time the failure situation will stabilize at a certain level of $\lambda(t) = C$.

Example

In a one year period four failures are found, so $\lambda = 4$/year and the R(one year) = $e^{-4/1.1}$ = only 1.8% or the chance that a system survives one year of operation without any stop through a failure is 1.8%, at $\lambda = 1$/year the R(one year) = 36.8% and only at $\lambda = 0.105$/year or one times in 9.5 years the R(one year) is 90% (0.90).

Exercise
What is the value of λ for R (one year) is 95% (0.95)?
Answer: $\lambda = 0.051$/year or one time in 19.60 years.

In the same way as for the degradation process, we can formulate the mean time between failures (= MTBF), the mean time to repair or eliminate a failure (= $MTTR_F$) and the amount of failures in a specified period of time, e.g. per year, (= λ_F), see figure 7.4.

The next expressions are possible:

For the availability due to unplanned maintenance generated by failures:
$A_F = MTBF / (MTBF + MTTR_F)$

For the number of stops for unplanned maintenance generated by failures:
$\lambda_F = MTBR / (MTBF + MTTR_F)$

For the total time to stop (TDT = Total Dead Time) due to unplanned repairs:
$TDT_F = \lambda_R \times (\lambda_F \times MTTR_F)$ in 8760 hours (one year)

TOT (= Total Operational Time = max. 8760 hours/year)

TOT_R (= Total Operational Time after repair) = TOT − TDT_R

TOT_F (= Total Operational Time after failure) = TPT_R − TDT_F

To improve the effectiveness of the maintenance activities originated by failures the following actions have to be taken. Make higher A_F, lower λ_F and lower TDT_F, or make higher MTBF and lower $MTTR_F$.

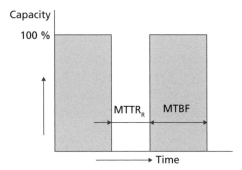

Capacity

100 %

MTTR$_R$ MTBF

Time

Figure 7.4

Example

Failures have taken place during the MTBR periods so if we look to the previous example we see MTBR =1035 hours and MTTR$_R$ =60 hours, λ_R = 8 and TPT$_R$ = 8280 hours.

If MTBF = 67 hours and MTTR$_F$ = 2 hours, then:

A_F = 67 / (67 + 2) = 67 / 69 = 0.971 = 97.1 %

λ_F = 1035 / (67 + 2) = 1035 / 69 = 15 times in 1035 hours or 8 × 15 = 120 times a year

TDT$_F$ = 8 × 15 × 2 = 240 hours/year

TOT$_F$ = TPT$_R$ − TDT$_F$ = 8280 − 240 = 8040 hours/year

Combined figures of A_R and A_F in A_T and λ_R and λ_F in λ_T

A_T = A_R × A_F = 0.945 × 0.971 = 0.918 = 91.8%

λ_T = λ_R + λ_R × λ_F = 8 + 8 × 15 = 8 + 120 = 128 times a year

In general, repair activities resulting from the unplanned failure process can not be planned and are being described as **Unplanned Failure Maintenance (UFM)**.

Remarks

- Sometimes it is also possible that unplanned break downs in production are occurring because of lack of orders, lack of raw materials, or materials of the wrong quality. Here we can also make use of a similar set of expressions, these figures play an important role by measuring the operational outcomes of industrial fixed assets like production plants, oil refineries, etc.

- In all cases we see in order to achieve improvements in A, λ, TDT and TPT, we have to increase the MTBR and the MTBF and lower the $MTTR_R$ and the $MTTR_F$.
- The availability A is a product of regular repair stops, unplanned failure repair stops and unplanned production break downs. This A is an important figure for the measurement of the efficiency of production plants.
- For production lines, A is usually calculated as the amount of hours we can actually produce in relation to the total available hours in unit time (e.g. one year). In case of fleet maintenance (e.g. ships, buses, aeroplanes, etc.) we use another calculation of A, in case of buses, the expression of A = amount of buses on the road / total amount of buses.

7.4 Maintenance Repair or Restoration Strategies

We have seen that when using or operating an asset, the effective output of the function of this asset decreases and at a certain point a repair activity has to take place to restore the effective output of this function. We called this process the **maintenance degradation process** and one of the products of this process is the generation of an amount of stoppage or non productive hours of the asset. In other words it gives availability through repair activities of A_R. The question is: how to control these stoppage hours? Another question is: what type of repair strategies are possible to restore the functions?

Before we can formulate repair strategies we have to go back to figure 7.1 of the degradation process where we saw a straight line from 100% effective output at time is zero to the limit line of x% at t_o. But in reality the degradation of an asset, or a part of the asset, is often not a straight line, instead there is a variation of values around this line. We have to know how wide this variation can be, in other words we have to know the behavior of an asset, or part of an asset, over time as a result of operating this asset.

Example
The wear of the tyre of a car depends heavily on the driver's style of driving, so the life time of a tyre can vary widely (10.000 to 35.000 miles).

Example
A population of light bulbs within a building has a mean life time of 1000 hours (from 970 to 1030 hours) with a relatively small variation around this mean life time (so has a small normal distribution).

The consequences of operating or using an asset, or part of an asset, have to be known before we can select an appropriate repair strategy. If we do not know this behavior over a period of time there is a problem that we have to tackle before we can choose the right strategy. In other words controlling the maintenance degradation process is controlling the *behavior* of the asset, or part of an asset, over a period of time.

When we look deeper in the wear or degradation process (see figure 7.5) than we see mostly assets, or parts of assets, with a mean life time and a well known normal distribution around this mean life time. Sometimes the variation is small (σ is small), for example the light bulbs, and sometimes the variation is wide (σ is great), see the tyre wear. This outcome of the form of the normal distribution has consequences on the choice of the repair strategy.

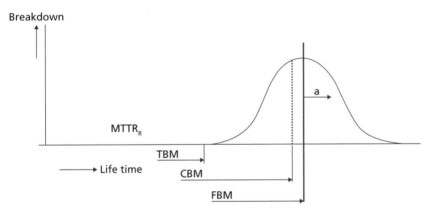

Figure 7.5 The wear and degradation process

7.4.1 Possible maintenance repair strategies

When we are looking to the degradation behavior of assets and parts seen over time, then we can generally define proactive strategies through three types of repair strategies and also the 'modification as improvement' activity. Also reactive strategies can be defined through repair after failure and modification as a result of improvement considerations.

Proactive Maintenance Strategy or Planned Preventive Maintenance:

- CBM = Condition-Based (planned by 'condition');
- TBM = Time-Based (planned by 'fixed' time);
- FBM = Failure-Based (planned by 'failure');
- MOD = Modification-Based (activated by 'improvement').

Reactive Maintenance Strategy or Unplanned Failure Maintenance:

- RFB = Repair after Failure Break Down (unplanned and not wanted by *break down* of the main function);
- FEM = Failure Eliminated by Modification (activated by *business improvement*).

For each type of repair strategy we will look at the considerations for choosing that specific type. The choice of the repair strategy will depend heavily on the wear behavior of an asset (or part) over time and also on the cost to execute the strategy. For failure repairs it will usually depend on construction, handling failures or quick changes in an operating environment.

I Condition-Based Maintenance

When will the **Condition-Based Maintenance (CBM)** be applied? CBM as a repair strategy will be used if the condition of the effective output of an asset, or part, seen over time can be measured, if there is a so called measurable variable, and if the output has a widespread variation in the normal distribution. So the *condition* is the guide for starting the repair activity (see figure 7.4)

Example

The car tyre has a well known relationship between tyre profile (groove depth) and potential danger through driving in heavy rain (aqua planing). So at a certain groove depth (2 mm) it is advisable to change or replace the tyre with a new one.

But in Egypt there is practically no rain over the year and we see cars on the street with no profile (down to the canvas!) on the tyres! And at Formula 1 (Grand Prix) races we often see the change of tyre types during the race depending on weather conditions, from so called slick tyres with no profile for dry weather to tyres with profiles for wet weather.

Example

Strike fighters of the Royal Dutch Air Force, often flying over the wet open sea, showed a totally different pattern of life time wear of certain bolts due to corrosion phenomena than American ones that were mostly flying over the dry desert areas.

These examples show another important phenomena of condition based maintenance, the change in behavior of the asset due to changes in the **operating context** in which the asset is used.

To apply CBM we have seen a need for *knowing the behavior* of an asset, or part, over a period of time by using it in a certain operating context and having the *possibility to measure* the *decrease pattern* of this behavior. Sometimes the cost of measurement is so high that another repair strategy has to be chosen.

CBM applies different types of measurement techniques, such as:

- vibration or sound techniques for analysing bearings;
- analysis of metal particles in lubrication oil (wear of a specific part!);
- lubrication oil analysis (general behavior of oil);
- inspection techniques (infra red, X-ray, etc.);
- inspection by ear, eye, feeling.

Another option to use CBM is provided by the new strong and intelligent devices like PLC (Programmable Logic Controller) and DCS (Distributed Control System). Boosted by chip developments, these devices offer new possibilities to store data and obtain condition information from it. When combined with telecoms technology it is possible that an asset can generate its own degradation pattern for certain functions, including issuing a warning when the limit of operation is reached. Equipment like pumps, compressors, valves, ventilators, etc. are increasingly controlled by certain process parameters. With pumps, for example, the flow control can give an indication of the wear of the impellor, beneath the well known bearing vibration control. With piston compressors the pressure and the temperature of the outlet can give an indication of imminent valve problems. For example, piston compressors of air conditioning systems in buildings are controlled from a distance and service mechanics on the road can immediately be guided to a possible compressor trouble. There is also the development of electronic garage equipment to control the status of the condition of several functions of a car visiting the garage.

Or look at the development with printers, whereby a printer can identify by itself the need for a new toner and is able to order a new one directly by telecoms.

These and other interesting examples show that a new era in maintaining equipment is coming. Equipment with intelligent electronic information systems will organize the need for repair by itself, for a specific condition and for time based maintenance activities.

2 Time-Based Maintenance

When should **Time-Based Maintenance (TBM)** be applied? TBM is applied as a repair strategy when the behavior of the degradation from a new or standard value to the allowed limit value shows a relatively small variation around a mean value. In other words the normal distribution curve is small and has a low value (see figure 7.5). The 'time' in this case can be a certain running time (an amount of hours), revolutions ($\times 10^6$) for bearings, weight, movements, etc. So a *fixed running* time is the guide for planning the repair activity.

An example of a TBM application is the change of the population of lamps in a building (e.g. X thousands), which are changed all at once at certain times after y burning hours.

Another example of TBM can be the changing of tyres at fixed mileage readings of city buses, which are all driven in more and less the same way.

In a hospital every year before the winter period begins, the V-belts of the ventilators placed on the roof are replaced by new ones. The life time of these V-belts is 1.0 to 1.5 years, so for security the maintenance department changes these V-Belts just before the time of break downs and just before the winter period, which always gives bad working conditions on the roof.

In the past TBM was known as preventive maintenance, which means the change of parts after fixed times without any relation to the real state of these parts. Nowadays, however, the term preventive maintenance has a wider meaning.

Remark: Choice of TBM versus CBM
Sometimes the costs of measurement are high and the choice of CBM is disputable, in which case the choice can be TBM. Looking at figure 7.5 we can see that CBM uses the mean life time of parts better than TBM. By applying CBM, the moment of changing a part can be determined accurately, and by applying TPM we know that the moment of changing depends on the shape of the normal distribution.

Example

TBM versus CBM
In a production line the net income per production hour is €100, the cost of changing a part is €150 and the time to repair (change of part) is two hours. In case of CBM in two years time three cycles of measurement and replacement take place and each cycle has two measurements, which cost each €75. In case of TBM in two years time four planned replacements will taken place with only the change of part required.

Which repair strategy will be the cheapest?

CBM: 3 × (2 × €100 + 2× €75 + 1× €150) = €1500
TBM: 4 × (2 × €100 + 1× €150) = €1400

Conclusion: TBM is in this example the best strategy.

Question
From what level of the net income per production hour will CBM will be the best strategy?
Answer: About €125/hour.

Example
Maintenance behavior of roll bearings
The behavior of roll bearings over time is a very special one. In principle the phenomena of the wear of a roll bearing is fatigue. Usually the life time of a roll bearing is calculated for 10^6 revolutions and during this period the degradation is very slow. But at a certain point due to fatigue particles breaking from the surface of the outer ring of the bearing where the power transfer from the ball(s) to the outer ring take place, this will result in a higher bearing temperature, a higher level of vibration and/or a higher level of noise. All these three phenomena can be used for condition measurements (and thus for a CBM strategy). After the first warning of the fatigue phenomena, special attention is needed because the bearing can suddenly be blocked totally with very often serious damage to the equipment as result. This means that the measurement strategy of this CBM case has to be organized in such a way that the critical point 'just' before the dangerous break down take place is measured! Then we have to stop the equipment and replace the bearing. So the life time of the roll bearing is characterized by a long period of time with a small degradation pattern and a short but heavy failure period with a great chance of significant damage if the equipment is not stopped in time.

3   Failure-Based Maintenance
When will **Failure-Based Maintenance (FBM)** be applied? FBM is used as a repair strategy when the result of the break down of the asset, or a part of this asset, has no influence on the main function, and the frequency of failure of this situation is not very high. FBM is also termed '*run to failure*'. So a '*failure*' in this case is the guide for planning the repair activity. This means also that the repair activity does not have to start immediately. The choice for this strategy is proactive.

Examples

> An example of FBM is the break down of one light bulb somewhere in a building and the replacement of this bulb at some point later by the caretaker.
>
> Or the break down of one of the four V-belts in the power drive of a ventilator. Once again the replacement can be done some time later, often with the replacement of the other three at the same time.
>
> Another example is the break down of a pump in a secondary system in a plant which can be handled under a FBM regime, e.g. in a cooling system.

Remark Applying FBM:

A special remark in terms of the application of FBM has to be made and this relates to exactly what can happen if this situation occurs. For a light bulb or one V-belt the situation is clear, a critical situation is not a realistic possibility. However, a pump break down can result in completely ruining the pump, which necessitates a high cost for replacement. In this type of situation further consideration should be given and another repair strategy chosen, e.g. CBM or TBM.

4 Modification (MOD) or Improvement-Based Maintenance

Modification (MOD) of a repair strategy is applied when it is possible by choosing another type of part (better quality bearing) or better quality of material (better corrosion behavior) to improve the time between repairs (MTBR). This result very often in less times to stop for repair (lower λ_R) , lower repair time, a higher A_R or a higher TPT_R and thus a higher production volume. The income from this extra production volume has to be so attractive that a good ROI is the result. This means that for this type of modification the business calculation often give the go ahead of the implementation.

Example

> By choosing a better quality bearing the MTBR could be extended from 8000 hours to 16000 hours, whilst the MTTR could also be decreased from 2 hours to 1 hour. The old bearing (1) will cost €100 and a new bearing (2) will cost €150. The replacement time is the same, one hour, and the net income per production hour is €75.
>
> For bearing 1 the total cost is: 2 times change of a bearing and two times 2 hours loss of production income: 2 × 100 + 2 × 2 × 75 = €500 (over 2 years).
>
> For bearing 2 the total cost is: 1 hour loss of production income: 1 × 150 + 1 ×1 × 75 = €225 (over 2 years).

5 Repair after Failure Breakdown (RFB)

Because a break down of main function(s) after failure(s) is unplanned and unwanted, there is no planned repair strategy like CBM, TBM and FBM. In nearly all cases the repair activities have to take place as soon as possible to rectify the failure troubles. We can call this: **Repair after Failure Break Down (RFB)**. Sometimes this repair activity is described as `corrective maintenance'. This is the major difference with the FBM strategy, where the result of a failure is known and the time to start the repair work does not have a great urgency.

Analyses of failure patterns indicate that about 50% of all unplanned failures are due to asset or equipment failures, like fatigue of shafts or bolts and break downs of computer boards. The other 50% come from human errors or failures, an operator uses the wrong valve or a mechanic leaves some tools in the machine, etc. Even when we organize all the repair work in a preventive way through CBM, TBM and FBM, 10 to 15% of all repair work will still be generated by break down failures.

6 Failure Elimination Modification

We have already seen that the best policy for rejecting this type of failure is to collect data about the different type of failures, construct a Pareto diagram of these different failures (see section 6.4.2), and modify the most ´expensive´ failures in such a way that the failure (type) is eliminated. Subsequently the effective operating time of the asset will increase, or A_F increases.

This process is also known as ´modification by eliminating failures´ (FEM activated by ´business improvement´). Because these kinds of failures are often widespread, e.g. for an asset more than twenty different type of failures are possible, the strategy to tackle the impact of failure processes is to collect and analyse the failure data. In addition to catalogue this data into different types of failure, count how often a specific type of failure happens and combine the results in a histogram. Usually we see the results back in the form of a Pareto diagram (see figure 7.6). Very often the outcomes are characterized by the so called 80/20 rule, which means that 20% of the types of failure result in 80% of the total effect of the failures. By concentrating on this 20%, the failures with the greatest effect are analyzed in terms of the possibilities for an improvement modification.

The modification with the highest Return on Investment (= ROI) or the shortest Pay Back Period (= PBP) is chosen to be implemented. This elimination of failures will continue as long as the results of the ROI or the PBP indicate opportunities for improvement activities.

This elimination of failures is the way to control and to improve the failure process of maintenance. It not only brings a higher level of reliability but also an improvement in production time, and consequently an improvement in production volumes, profits etc.

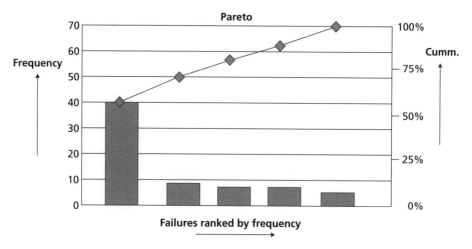

Figure 7.6 Example of Pareto diagram

7.4.2 Case studies

Mix of maintenance repair strategies
We have seen that repair strategies consist of proactive preventive strategies based on planning (CBM, TBM and FBM) and of reactive non-preventive failures, not based on planning (RFB). In a well organized situation the percentage of repair work covered by preventive strategies reaches the 90% mark (80% to 90%), whilst the rest of the repair work accounts for about 10% to 15% and still consists of unplanned failures. The maintenance work undertaken as part of CBM, TBM and FBM is dependent upon the type of equipment and the operating context of this equipment. Every equipment (asset) working in a specific operating context will have its own mix of maintenance repair strategies. In section 7.4 we will see how this mix can be developed in practice.

Maintenance repair strategy scheme
In the maintenance repair strategy scheme an overview is given of maintenance repair strategies and some of the main aspects that will play an important role in selecting a strategy.

Asset Behavior

Preventive: CBM: pattern known   wide normal distribution   measurable

                        TBM: pattern known   narrow normal distribution fixed time

                        FBM: pattern known   low risk profile   repair after failure

                        MOD: modification for improvement of maintenance needs

                        (ROI or PBP)

Not Preventive: RFB: pattern unknown unplanned and unwanted   repair after failure

                        FEM: modification for improvement of reliability (ROI or PBP)

7.4.3 Examples of typical maintenance strategies

Typical CBM examples are:
- rotating equipment of all types;
- filters;
- corrosion in pipelines;
- erosion phenomena (dredge ships) in pumps, valves and pipelines.

Typical TBM examples are:

- worn parts such as liners, bearings, valves, filters, corrosion in pipes.

Typical FBM examples are:

- failure of a V-belt (one of six parallel);
- break down of an unimportant pump.

Typical RFB examples are:

- shaft broken by fatigue phenomena;
- tool forgotten and left in machine, resulting in a lot of damage;
- operator reacting too late to a warning signal, pump breaking down.

7.5   Maintenance Strategies for New Products

In 'tools for selecting maintenance strategies' we explained that in order to be able to select the right maintenance repair strategy, it is essential to have knowledge of the degradation behavior of an asset or parts of this asset (the wear patterns) within the operating context. Also the limit on the level of wear prior to exchanging assets or parts is very important, because this limit gives the amount of changes in one year (λ_R) and has a direct influence on the availability (A_R) of an asset.

Now we want to know how to develop an overview of the total potential 'need for maintenance'. This need for maintenance is the total number of man hours we have to spend in maintaining the function (specifications) to the level that the operator requires to operate. We will apply the Reliability Centred Maintenance (RCM) methodology for generating this need. Another important factor is the relationship between the asset operation and the goals of the company, as these have to be in line.

In addition to RCM, some techniques are described (FME(C)A, FCRA) for looking in detail at those situations that are critical and which make analyses for improvement possible.

Through utilising the tools and techniques for selecting maintenance repair strategies (maintenance tasks) a well formulated maintenance policy in relation to the company's goals can be formulated.

7.5.1 Functional critical Reliability Centred Maintenance (Fc-RCM)

The **Reliability Centred Maintenance (RCM)** methodology originates from the USA and was developed during the late 1970's and early 1980's. RCM was developed from the insight that for big complicated and dangerous systems, like nuclear power plants and big aeroplanes like the Boeing 747, the more common maintenance strategies such as break down (corrective maintenance) and the so called preventive maintenance were insufficient to handle the maintenance practice and deliver a really practical and safe maintenance strategy and maintenance concept. RCM delivers on basis of failure modes and failure effect analyses and an overview of the criticality of systems, sub systems and parts. It is also an excellent methodology for selecting the right maintenance repair strategy.

RCM is: *'a process used to be determine what must be done to ensure that any physical asset continues to do what ever the users want it to do in its present operating context.'* So for a physical asset the users have to determine what it has to do and if it is capable of doing this. Therefore we have to firstly define the function of each asset (in its operating context), the functional specifications (desired standards of performance) and possible solutions of assets that can fulfill these functions and specifications. In case of an industrial installation a functional break down will be made in the form of a functional decomposition in which, under the main function, a lot of sub functions arranged in different layers can be determined.
Before we can start the RCM process users have to know two things about the applied assets:

- determine what they want to be done;
- ensure that the applied asset is capable of doing what is wanted from the outset.

So users have to be clear about their wishes in terms of what is wanted in a specific operational context. Functions and desired standards of performance (functional specifications) are used to specify the possible objects or assets. Often more than one solution to the function-functional specification set is possible.

Example

> function: generate electrical power
> functional specifications: alternated current, 50 Hz, 230 V
> (function solution(s): steam-turbine generator, windmill, PV-system, etc.

Usually the functional specifications will provide the final choice of a solution. In figure 7.7 a model is shown where one picture includes a complete function statement. This is the Hamburger model. The model is also the basis for a functional decomposition which will be used by developing a Functional critical-Failure Mode and Effect Analysis (*Fc*-FMEA).

Remark: A function statement should consist of a verb, an object (asset) and a desired standard of performance (specifications).´

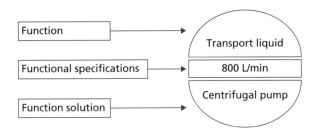

Figure 7.7 Hamburger model

Functions can be split into two categories:

- primary functions cover issues such as: production speed, output, product quality;
- secondary functions cover areas such as: safety, containment, comfort, structural integrity, economy, efficiency of operation, environmental regulations.

To achieve a stable operation, the initial capability of an asset (*´what it can do´*) has to be higher than the desired performance (*´what its users want it to do´*). In the case of machinery where a decrease in performance as a consequence of the wear phenomena is normal, an initial capacity has to be chosen which is 20% to 25% higher than the desired performance (see figure 7.8). Very often schemes of production installations are used at the start of the RCM process.

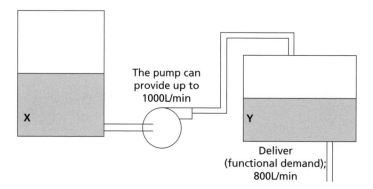

Figure 7.8 Chosen performance (1000 L/min) of a pumping system is 20% higher than the desired performance (800 L/min)

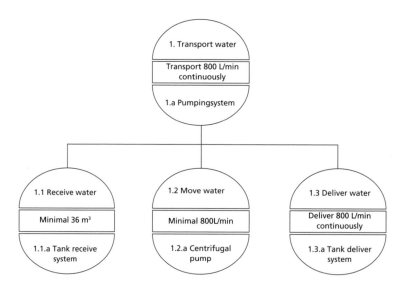

Figure 7.9 Functional decomposition (Hamburger model) of a pumping system

We will now describe how to make a functional decomposition of a pumping system (figure 7.9). See also chapter 3 on how to make a Hamburger model.

7.5.2 How to make a functional decomposition?

We start by formulating the main function (see figure 7.9); in this case it is 'to transport water' (at least 800 litres/min). The chosen solution (an object that can fulfill this function) is a pumping system. The decomposition starts by investigating and formulating the underlying sub functions, here: 'receive water', 'move water', 'deliver water' with the solutions: 'tank receive system', 'centrifugal pump (system)' and 'deliver water`. We stop this process when we reach a sub function (and solution) where we can directly formulate a repair activity or maintenance task.

7.5.3 Functional failures

If for any reason the asset is unable to do what user wants, so that it cannot fulfill the functional specifications, than it is considered to have failed.

Definition: 'Failure is defined as the inability of any asset to do what its users want it to do'. Failures can be defined more precisely (capable of being measured) by using the functional specification as the boundary between 'operable and fail state'. So a functional failure can be defined as follows: 'The inability of any asset to fulfill a function to a desired standard of performance'.

This means that when a piece of equipment goes below the desired standard of performance, it can still run or operate. This event often gives rise to major disputes between the technical maintenance department and the production department.

Example

(See figure 7.8.) The pump is not in a state of functional failure if the output is above 800 L/min. However, when the pump suffers from wear and drops to a lower level, say 780 L/min, then it is in the state of functional failure and has to be repaired or replaced. So the repair activity will be activated by the levels of functional failures. If the production output rises from 800 to 900 L/min, then the repair activity has to start at 900 L/min level, the new level of the functional failure! The maintenance department now has to repair this pump at an increased frequency, e.g. twice as many times as in the old situation!

The product development team has to realize that the real operating context of an object will dictate the real maintenance plan.

Remarks
- Equipment like pumps, centrifuges, filters and compressors that can be characterized by decreasing output though the wear phenomena, often have different level of desired performance during operation through changes in the production levels.
- Maintenance departments have to fulfill these changing situations and must follow them. This means having to be flexible in arranging and planning repair work. This is often not the case, resulting in troublesome relations between the maintenance and the production departments.
- Clear definitions of functional failures as a key starting point for repair activities can overcome these troubles.

7.6   Functional critical Failure Modes and Effects Analysis (*Fc*-FMEA)

The functional statement and the functional failures give outlines about the desired performances but not the reasons for the failures. Now we have to identify the *failure modes* which are reasonable likely to cause each functional failure and the *failure effects* associated with each failure mode. This is done by performing a **Functional critical Failure Modes and Effects Analysis (*Fc*-FMEA)** for each functional failure.

The uniqueness of the *Fc*-FMEA is that at each stage of the functional decomposition it is asked how critical the functional failure of this function is in relation to the main function, in terms of the complete system (object) going down.

Definition: *A failure mode is any event which causes a functional failure.*

Categories of failure modes are:

- falling capacity;
- increase in desired performance (capacity or increase in applied stress);
- initial incapability.

How much detail?
Failure modes should be defined in enough detail for it to be possible to select a suitable failure management policy (see also figure 7.9 the functional decomposition). When listing failure modes, do not try to list every single failure possibility regardless of its likelihood.

Definition: *Failure effects describe what happens when a failure occurs.*

Example

Fc-FMEA

Function:1
To transfer water from tank X to tank Y at not less than 800 L/min.

Function solution: A
Centrifugal pump.

Functional Failure: 1
Transfers less than 800 L/min.

Failure Modes:	Failure Effects:	Failure Consequences:
1. Impellor worn	less throughput	less critical
2. Mechanical seal leaks	less effective throughput	less critical
3. Partially blocked suction line	less throughput	less critical
4. Power break down	no throughput	less critical
Etc.		

Remarks

- The functional failures can be characterized from 'not critical' up to 'very critical' (see figure 7.10 Criticality Matrix). Depending on the severity of the outcomes, a repair strategy can be determined. This will vary from choosing a Failure Based Maintenance (FBM) approach, via a task selection methodology to a full FMEA approach.
- The Criticality Matrix can be applied when no real failure data is available. This is mainly the case for industrial installations. A small group of in-house experts can each make an estimate of the possible frequency and severity of the functional failure and give their so-called expert interpretation.
- The Criticality Matrix is also helpful in developing a FME(C)A, when data is not available. Normally the C of the FME(C)A is extracted from failure data, but this are often not available. In such cases a lot of firms use an extended criticality matrix with their own failure consequences in the form of lost income, injuries, fatalities, plant damages, environmental damages, etc.

7.6.1 Failure consequences

The consequences of a functional failure differ, with a greater or lesser impact on the people involved, the installation and the surroundings. Usually the following division of failure consequences is applied:

- Hidden failures:
    - multiple failure consequences.
- Evident failures:
    - safety and environmental consequences;
    - operational consequences;
    - non-operational consequences.

Definitions:
1   *A hidden function is one whose failure will not become evident to the operating staff under normal circumstances if it occurs on its own.*

2   *An evident function is one whose failure will on its own eventually and inevitably become evident to the operating staff under normal circumstances.*

With the Failure Consequences Analysis (FCA) diagram (see figure 7.11) it is possible through a series of questions to categorize the failures. The outcomes will be a repair strategy or a modification.

Special attention must be paid to the *hidden failure*. This type of failure requires a special repair strategy: 'failure finding' task, which will have to address each specific hidden failure (a form of inspective maintenance, with periodic controls, part of CBM or even modification).

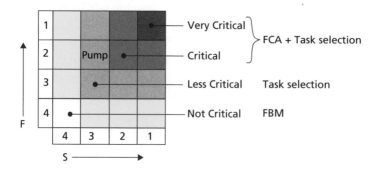

Figure 7.10 Criticality Matrix. FS1.1 is very critical and FS4.4 is not at all critical.

Ranking of failure consequences

We can rank the failure consequences as follows:

Ranking	Maintenance Repair Strategy
• *Not critical at all*	
FS4.4	Direct choice FBM
• *Not critical*	
FS4.3, 4.2, 4.1, 3.4, 2.4, 1.4	Task Selection (CBM, TBM, FBM)
• *Less critical*	
FS3.3, 3.2, 3.1, 2.3, 1.3	Task Selection (CBM, TBM, FBM)
• *Critical*	
FS2.2, 2.1, 1.2	FCA and Task Selection
• *Very critical*	
FS1.1	FCA, Task Selection or Modify

With the help of the Failure Consequences Analysis (FCA) scheme we can determine which categories of failure consequences are involved (see figure 7.11)

The Failure Consequences Analysis (FCA) is a part of the *Fc*-RCM process and will be applied in the *Fc*-FMEA scheme.

A: safety problem
B: environment problem
C: production problem
D: not serious production problem
F: hidden failure problem

Other questions:(1) is maintenance behavior known and/or (2) is operating context known?

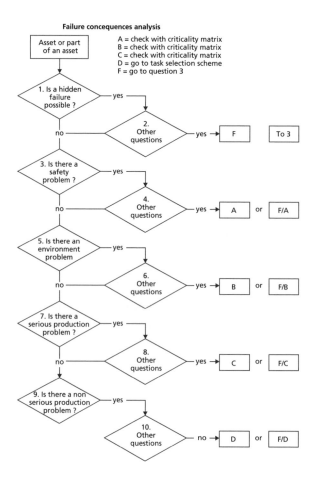

Figure 7.11 Failure Consequences Analysis (FCA) scheme

Task Selection Scheme

Remark

In a normal situation with a proven design, a choice of FBM, TBM or CBM is always possible. Only in special circumstances, or a completely new design modification, will reconsideration of the design criteria occur.

Selecting a maintenance strategy is sometimes indicated by different authors as 'task selection' and FBM, TBM and CBM are named as maintenance tasks (see figure 7.12)

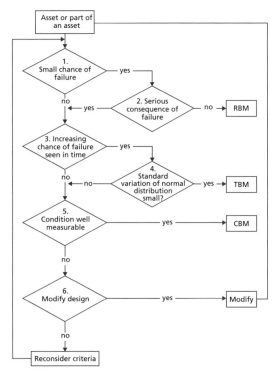

Figure 7.12 Task Selection Scheme

Example

Task Selection Scheme

Asset or Part	Ques. 1	Ques. 2	Ques. 3	Ques. 4	Ques. 5	Ques. 6	Repair strategy
Pump	Yes	Yes	Yes	Yes	-	-	TBM
Bearing	Yes	Yes	Yes	No	Yes	-	CBM
Centr.Compressor							

Exercise

Fill in the scheme for a centrifugal air compressor (in an air conditioning unit).

Answer: Normal CBM (TBM is possible).

7.6.2 Development of the full *Fc*-RCM process

The development of the **Fc-RCM process** will be undertaken by applying the pumping system example. It consists of the following seven steps:

Step 1: Describe the system and the boundaries of the system: *pumping system*
Step 2: What are the main function(s) of the system: *transport water, at least 800 L/min.*
Step 3: Make a functional decomposition of system. (See figure 7.9, Hamburger model.)
Step 4: Make a list of all functions and sub functions, functional specifications of these functions and the functional failures.
Step 5: Make a FMEA with all the functions (Fu , solutions (SO), functional specifications (FS), functional failures, possible failure modes (FM) and failure effects (FE), classify with the criticality matrix (CM) the failure consequences for each functional failure (FF), see below the so called *'functional critically* FMEA' or (*Fc*-FMEA) for the transport function of water by a centrifugal pump (see table 7.1). This is classified FS2.3, which means *'less critical'* and the ranking gives the outcome *task selection'*.

FU	SO	FS	FF	CM	Ranking	FM	FE
1	A	1	A	1		1	
Transport Water	Centr. pump	Q = 10 m³/hr and h = 100	Q< 10 and or h< 100	FS2.3	Task S.	Impellor wears	Less throughput
						2	
						Seal leakage	Less throughput
			B	2		1	
			Q=0, h=0	FS2.3	Task S.	No power	No throughput
						2	
						Blockade of Surge pipeline	No throughput.
						3	
						Bearing damage	No throughput

Table 7.1 Fc-FMEA

Step 6: Outcome of CM choice and ranking is task selection or FCA analysis.
If the pump of step 5 has been classified FS2.2 than the ranking is *'critical'* and a FCA analysis has to be executed. For the results see table 7.2 'FCA Analysis'.

Part	Qs.1	Qs. 2	Qs. 3	Qs. 4	Qs. 5	Qs. 6	Qs .7	Qs.8	Qs.9	Qs.10	Outcome
Impellor	No	-	No	-	No	-	Yes	Yes	-	-	C and CM
Mech. Seal	No	-	No	-	No	-	Yes	Yes	-	-	C and CM
No power	No	-	No	-	No	-	Yes	Yes	-	-	C and CM
Blockade	No	-	No	-	No	-	Yes	Yes	-	-	C and CM
Bearing	Yes	Yes	No	-	No	-	Yes	Yes	-	-	F/C and CM

Table 7.2 FCA Analysis

CM analysis is now used again for further investigation:

Part	Ranking	Result	Effect	Measurements
Impellor	FS2	Not Critical	Task selection	No
Mech. Seal	FS2	Not Critical	Task selection	No
No power	FS1.1	Alarm	Modification	Double power system, task selection
Blockade	FS2.1	Critical	Incident	No
Bearing	FS2.1	Critical	Modification	Double pump, task selection

Develop a maintenance plan on basis of the results of 'task selection' (table 7.3).

Asset or Part	Ques. 1	Ques. 2	Ques. 3	Ques. 4	Ques. 5	Ques. 6	Repair strategy
Impellor	Yes	Yes	Yes	No	Yes	-	CBM
Mech. Seal	Yes	Yes	Yes	No	Yes	-	CBM
No power	Yes	Yes	Yes	No	Yes	Yes	No strategy, incident
Blockade	Yes	Yes	Yes	No	Yes	Yes	No Strategy, incident
Bearing	Yes	Yes	Yes	No	Yes	-	CBM

Table 7.3 Task selection

Step 7: Maintenance plan
A maintenance plan can be developed from the outcomes of the task selection and task specifications input. The plan can be as follows:

FM	Repair Strategy	Frequency	By whom Hours	New part / Unique nr: price
Impellor wear	CBM	per two years	Mechanic (1x4)	Impellor / 100.0025: €2000
Seal leaks	CBM	per one year	Mechanic (2x2)	Seal / 100.0273: €750
No power	SOO	unknown	E-mechanic	
Blockade	SOO	unknown	Mechanic	
Bearing wear	CBM	per two years	Mechanic (1x3)	Bearing / 100.3698: €450
Total:			10 hours (in 2 years)	

With the total man hours and the unit prices of the parts we can now calculate the total maintenance costs.

Note: The total of man hours we called the 'maintenance need' is generated by the product, installation, etc.

7.6.3 What Fc-RCM achieves

When applying the Fc-RCM methodology we can achieve the following points:

- The development of a maintenance plan on the basis of the functional specifications of the assets to maintain.
- The development of a Fc-FMEA scheme with all the possible functional failures, failure modes and effects.
- Having an overview of all critical functions and functional failures and a ranking of the severity of these critical functions.
- A list of weak points of the assets and indications for improvement or even modification.
- Revised operating procedures for the operators of the asset.
- Having an overview of all planned maintenance work (hours, who is doing it, price of parts), in other words: the total ´maintenance need´.
- Having the use of a maintenance plan as a basis for improving.
- Having the use of multidisciplinary teams, that can execute the RCM methodology.

Remark

The introduction of the Fc-RCM methodology in an organization needs the employment of small teams that include at least one person from the maintenance department and one person from the operations department. These teams, supervised by a RCM facilitator, will execute the RCM review process. These RCM teams will consist of 4 to 6 persons from different departments.

7.7   Product Development and Maintenance Plan

On the basis of the *Fc*-RCM methodology a product development team, in particular the after sales member of the team, can develop a basic maintenance plan. It will be very useful for manufacturers to use the *Fc*-RCM methodology as the basis for delivering the (standard) maintenance plan as a basic plan for the operation. It has to be part of the PSPDoc.

In reality the product can be applied in very different circumstances, the well known *operating context*. This means that under different operating circumstances the behavior of wear and tear on parts can be completely different, and so is the total amount of maintenance needed as a consequence.

Examples

Aeroplanes flying over the sea are sensitive to crack corrosion through chloride ions and require need extra inspections and more frequent repairs than the same aeroplanes flying mostly over land or deserts.

Ships' engines running on gas oil have a totally different picture in terms of wear on parts such as cylinder liners, piston rings, valves, etc. than those burning heavy diesel fuel. But the price difference between gas oil and heavy fuel is so great that operating with heavy fuel always 'wins', despite the higher maintenance cost associated with applying this cheaper fuel.

In practice most manufacturers do not know exactly what the operating contexts of their equipment are, so real knowledge can only be gained by contacting a range of clients and drawing upon their experiences for product improvement programs.

There are manufacturers of capital industrial goods that deliver a long life product support and service contract. They are in the position to collect data from all types of operating contexts and develop maintenance plans appropriate to these different situations.

7.8   Commissioning and Early Equipment Management

Commissioning and **Early Equipment Management** marks the testing phase after on site installation of the full equipment for the production system or heating system. Early Equipment Management originates from the Japanese methodology of the Total Productive Maintenance (TPM). During this phase the equipment has to prove that it fulfills the requirements of the client. The starting up period is also part of this phase.

7.8.1 Commissioning of an installation

Before the hand over of a new installation of product or building takes place, the commissioning phase starts with formal measurements to prove that the installation works properly in line with the (ordered) client requirements. Usually a special test protocol is applied. Sometimes the commissioning and hand over takes place in the manufacturer's workshop, but usually it happens on site as the initial phase for the starting up of a new factory. The Hamburger model contains all the information of the functions involved in the functional specifications and the solution and is, therefore, a very helpful tool to execute the commissioning. This phase ends by composing a commissioning report.

7.8.2 Early Equipment Management

The Japanese TPM philosophy (see also section 7.8) states that the operation always must start with equipment that from the very outset completely fulfills the client requirements and that improvement activities designed to make the installation 'better' also start at this point.

No deviations from the standards of the requirements are allowed. So information is also collected about the operational and maintenance behavior to make these improvement activities possible.

The knowledge and experiences collected during the operation and maintenance of an installation can be very useful for the design of new installations or the modification of existing ones. Also knowledge and experiences gathered during the design, construction and commissioning of an installation are very useful in optimizing the operation and maintenance of an installation. TPM pays a lot of attention to flows of both knowledge and experiences. In addition, the personnel involved are encouraged to work on early equipment flows (see figure 7.13).

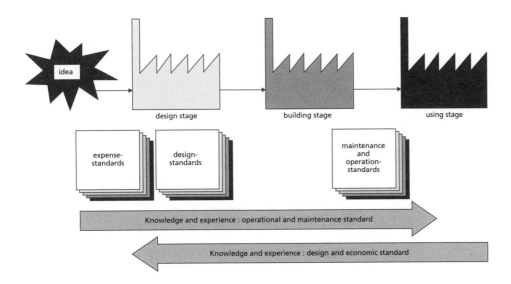

Figure 7.13 Early equipment flow

7.9 Total Productive Maintenance (TPM)

TPM originates from Japan (around 1980) and states that for productive systems (plants, trains, aeroplanes, etc.) the outcomes of the productivity of a system are a result of all the stakeholders in a company. From top management to the work floor, everybody is involved with the process of TPM, speaking the same language and working on continuous improvement of the productivity. Generally speaking this methodology can be seen as an operational follow-up to the IDE ideas in an operational production environment.

Definition of TPM is a follows: *A process in a company that enables the productivity of the production assets to be continually and systematically improved.*

An essential for TPM is the creation of self supporting multi disciplinary teams around the production lines which are responsible for the total production effort, guided by performance indicators. The teams can make autonomous decisions within the confines of the performance indicators. All the other departments of the organization have to support these teams. This means a completely different view on organization than is usual in the more traditional hierarchic organization.

Central to TPM is knowledge and information around the production equipment, the effective figures, the organization of the teams, especially for maintenance, the so called **autonomous maintenance** and the improvement efforts of the teams (see figure 7.14).

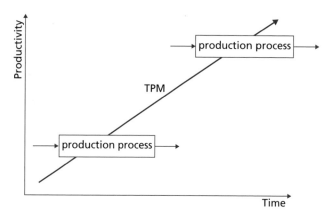

Figure 7.14

TPM recognises that the main stakeholders in the production, maintenance and engineering departments have to work together in an integrated way by analyzing problems, by thinking in terms of possible solutions, by providing input into design possibilities and by commissioning and starting up new installations (see figure 7.15).

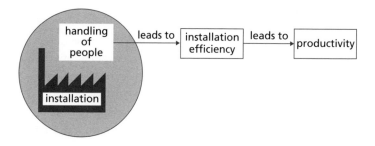

Figure 7.15

TPM is an overall system for organizing continuous improvement, consisting of seven main pillars:

1. Overall Equipment Efficiency (equipment and plant improvement).
2. Eliminating the 'six losses'.
3. Consolidating results of improvement:
    -   autonomous maintenance;
    -   planned maintenance.
4. Early equipment maintenance.
5. Leadership and motivation (how to organize TPM).
6. Dividing of production equipment.
7. Education and training.

We only discuss the first two pillars of TPM. The other mentioned main pillars of TPM we do not consider as a subject for this book

7.9.1   Overall Equipment Efficiency (OEE)

The basic measure associated with TPM is **Overall Equipment Efficiency**. OEE is a simple metric that highlights the status of a manufacturing process and is based on three (effective factors of) main losses (see figure 7.16):

- A = Availability (time losses). The share of the actual production time and the planned production time. Downtime: planned (preventive maintenance) or unplanned (break down) stops will reduce the availability, but also set-up times, adjustments, lack of operators (vacation, holidays) and lack of orders.
- P = Production Losses or Production Efficiency. Loss of production for example due to under-utilization of the machinery, reduced production speed or idling.
- Q = Quality Losses or Quality Efficiency. The amount of the production that has to be discharged or scrapped (defects and rework).

Each loss factor can be calculated quite simply:

Availability (A) = (planned production time – downtime) / (planned production time)

Productivity (P) = (cycle time × number of products processed) / (production time)

(The production time = planned production time – downtime)

Quality (Q) = (number of products produced – number of products rejected) / (number of products produced)

We can calculate the OEE - or put it differently: the *useful operational time* (see left corner figure 7.16) - with the formula:
OEE = Availability (%) × Productivity (%) × Quality (%)

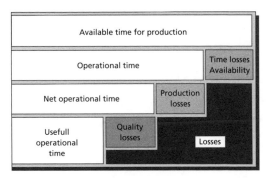

Figure 7.16 Time losses, production losses and quality losses

Example

What value can we expect for the OEE?

For a factory production line, the analysis of the OEE gives the following figures:
A = 0.90; P = 0.85; Q = 0.75, so the OEE = 0.57 = 57.0%

Is this outcome for OEE a normal result? Yes this is a very common figure when starting an OEE analysis (figures between 0.45 and 0.60 are often found).
If A = 0.90; P = 0.90 and Q = 0.90, then the OEE = 0.729 = 72.9%

Example

Calculating an OEE

For an installation the following data is measured for week x:

* total time for production: 80 hours
* registered losses in time: 12 hours
* total production: 10 tons
* rejected production: 1.2 tons
* standard capacity: 0.20 tons/hour
* set up capacity: 0.16 tons/hour

Calculation OEE:
Availability A = (80 − 12)/80 = 0.85

Productivity P here is: P= net capacity / standard capacity = (net capacity/set up capacity) × (set up capacity/standard capacity)
Net capacity = 10/(80 − 12) = 0.147 tons/hr
So P = 0,147/0.16 × 0.16/0.2 = 0.74

Quality Q = (10 −1.2)/10 = 0.88

OEE = A × P × Q = 0.85 × 0.74 × 0.88 = 0.55

Remark

The difference between the ideal OEE which is 100% and the realized OEE is called: 'the hidden factory'. The hidden factory in the example above: OEE_{hidden} = 100 − 55 = 45%.

7.9.2 The six big losses

Generally speaking there are **six big losses**, which can be combined into the three main losses (A, P and Q), see figure 7.17. Each of the mentioned losses can be a subject for work on improvements, the continuous improvement process. They are mostly executed by small teams as part of an improvement program (innovation agenda). The six big losses are combined with the three main losses: A, P and Q of the OEE.

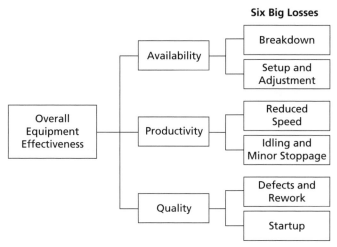

Figure 7.17 Six big losses

7.9.3 The other main pillars of TPM

Consolidating results of improvement
TPM is always practised by small groups or teams around the main process for executing maintenance. The way of operating is autonomous and planned maintenance:

- Autonomous maintenance
Autonomous maintenance is performed by a small team that daily operates the main process and that controls the behavior of this process in such a way that the OEE figures will be realized. So this group is autonomous to execute the daily maintenance (repair) activities and failures, eventually with extern help, if they decide to ask for an extra working force.
- Planned maintenance
All the other regular maintenance repair work is planned and guided by standards and also here improvement activities belongs to TPM.

Early equipment maintenance (see section 7.7.2)

Leadership and motivation (how to organize TPM)
TPM can only exist when the leadership is based on integrated team work, horizontal or process oriented and that stimulates continuous improvement.

Dividing of production equipment
Conform the ideas about functional decomposition, the total of processes (or the complete factory) will be split up in functional parts in such a way that small groups (consisting of operators, maintenance mechanics, quality personal) can rule the daily operation (guided or controlled by OEE figures).

Education and training
Permanent training on process knowledge, improvement techniques and social team behavior are taken place. These are completely in line with the IDE philosophy.

Summary

In this chapter the origin of the maintenance need is explained by introducing the degradation process and the failure process of course in combination with the operating context. This maintenance need is expressed in hours. Both processes have their own optimal policy to decrease the hours of the maintenance need, which will result in more productive hours of a machine or installation. So this outcome means a higher Availability. By means of a special *Fc*-RCM we can translate the maintenance need into a sound maintenance concept. For manufacturers of products, this *Fc*-RCM methodology provides a good opportunity to deliver a real basic maintenance concept for their future clients. We finish this chapter with a short overview of the TPM methodology as a driver for improvement activities during the operational phase of the product or installation.

Exercises

1. Calculate the A_{TOT}, λ_{TOT}, TDT_{TOT} and TPT_{TOT} for the following situation:

TOT = 8760 hours/year
MTBR = 1700 hours MTTR$_R$ = 52 hours
MTBF = 170 hours MTTR$_F$ = 3 hours
Give some suggestions for improvement.

2. Select a maintenance repair strategy for the following parts:

- cylinder liner of a piston compressor;
- roll bearing of a centrifugal compressor;
- filter set for filtering waste water;
- a lot of lamps in a big office;
- tyres of buses in a transport company.

3. On a production line the net income per production hour is €175, the cost of changing a part is €100 and the time to repair (change of part) is 2 hours. In case of CBM, in three years time five cycles of measurement and replacement take place and each cycle has three measurements, each of which cost €50. In case of TBM in three years time seven planned replacements will taken place with only the change of part required.

Which repair strategy is the cheapest?

4. *Fc*-RCM

In an installation a centrifugal air compressor is mounted.

- Develop a *Fc*-FMEA for this air compressor.
- If there is a not critical situation for this compressor, what is a suitable maintenance repair strategy for the compressors´ impellor?
- Develop a maintenance plan for this compressor.

5. TPM

For an installation the following data is measured for week x:

- total time for production: 120 hours
- registered losses in time: 15 hours
- total production: 21.0 tons
- rejected production: 2.0 tons
- standard capacity: 0.25 tons/hour
- set up capacity: 0.22 tons/hour

Calculate the OEE

- Availability A = ...

Productivity P = net capacity / standard capacity = (net capacity/set up capacity) × (set up capacity/standard capacity)

- Net capacity = ...

- Productivity P = ...

- Quality Q = ...

  OEE = A × P × Q = ...

Literature

- Wireman, T., *Total Productive Maintenance*, Industrial Press, 2004, ISBN 0831131721.
- Smith, A.M. and G.R. Hinchcliffe, *RCM*, Elsevier Butterworth-Heinemann, 2004, ISBN 0471591327.
- Moubray, J., *RCM II*, Butterworth-Heinemann, 1999, ISBN 0750633581.
- Andrews, J.D. and T.R. Moss, *Reliability and Risk Assessment*, 2nd edition, Professional Engineering Publishing, 2002, ISBN 1860582907.
- Rummler, G.R. and A.L. Brache, *Improving Performance*, Jossey-Bass Publishers, 1995, ISBN 0787900907.
- Zaal, T.M.E., *Profitable Driven Maintenance*, 2008, Lecture notes Hogeschool Utrecht.
- Kelly, A., *Maintenance Planning and Control*, Butterworth, 1986, ISBN 0408030305.
- Blanchard, B.S. et.al., *Maintainability*, John Wiley and Sons, 1995, ISBN 0471591327.

Chapter 8
Answers to the problems

Chapter 1

1. The *pro´s* of the IDE process:

- working on basis of real client wishes;
- it gives good insight in the real cost price and future profits;
- the ICT systems can be filled with the best structured models so reuse of knowledge is possible;
- working with multidisciplinary teams applying the concurrent engineering principles gives better products that can be manufactured very quickly.

The *con´s* can be:

- it is not possible to implement IDE 'the half way', because IDE is special way of thinking and working. It means that often the organization must be turned from vertical oriented (hierarchic) to horizontal oriented (client);
- an organization can be hierarchic oriented so strong, that is not able to fulfill the IDE philosophy;
- for building a real structured engineering data base much effort is necessary, and it is very expensive;
- teams working over boundaries of departments are not in the interest of these departments.

2. The main IDE competences are:

- client oriented: can translate client wishes in a functional way;
- design process is guided by Life Cycle Engineering principles, can handle these principles (including the Life Cycle Costs);
- can make functional decompositions of products;

- can make structured ICT models on basis of these functional decompositions for storing and reuse of this knowledge;
- can work in multidisciplinary teams as team player with a sound social behavior.

3.

The six IDE conditions are (mark):	0 – 35%	35 – 65%	65 – 100%	
• organization horizontal client and process oriented:				
• clear goals (e.g. for product development):				
• organization drivers for innovation and continuous improvement (changes, teams):				
• use of ICT technology as a real business process for the storage and reuse of knowledge:				
• working with IDE principles:				
• personnel who can handle and execute the IDE processes and will, therefore, have the right IDE competences:				
Total				$\sum = ...$

Each condition must have been marked for at least 60% and the mean value of the six has to be minimal 65%.

Note: A mean value below 35% means that an organization is unable to work with the IDE philosophy. A score above 65% indicates that the organization can fulfill the conditions to apply the real IDE philosophy.

4. The cash flow is $= 375 - 300 = €75$ million/year.
Depreciation is $= 10\%$ of $300 = €30$ million/year.
So the profit is cash flow minus deprecation $= 75 - 30 = €45$ million/year and the ROI is $= (45/300) \times 100 = 15\%$.

5. $0 = -25 + (1 - 0.4) \times \text{Turnover}_{crit}$

$\text{Turnover}_{crit} = 25/0,6 = €41,7$ million/year.
This means that above the value of this turnover we have the situation of profit, and below this turnover value there is the situation of losses.

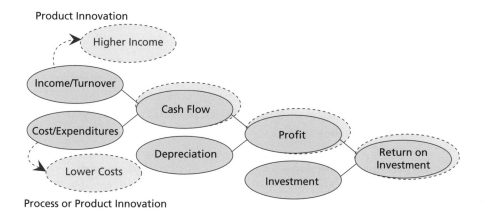

Product Innovation

Process or Product Innovation

Figure 8.1

6. As manufacturer you can plan an innovation agenda with the next main topics:

- Improvements of existing machine guide by client wishes like: higher production outputs, or lower energy consumption or better handling opportunities for operators and maintenance technicians.
- Improvements by widening the assortment for serving new clients or new markets.
- Improvements for lowering the production costs by design optimizing or introducing new production methods.

At least you ought to look or study the possibility for complete new products (the most risky way) to enter new markets or improve the existing technology.
All these improvements can lead to higher turnover and or lower costs so better profits and hopefully better ROI's.

Extra Exercises
1. What means the introduction of the IDE philosophy for the way of working in the building environment?
2. What type of organization is the most efficient type for applying IDE?
3. What can the IDE philosophy (way of thinking way of working) bring to your working environment ?

Chapter 2

1. Teams of students for engineering project work will not consist of more than five persons (four is even better). Each team member has to play a specific role e.g. coordinator, designer, marketer, production engineer, product support specialist. Each team member makes its own Belbin profile and will also report about its experience with the role to be played.

2. For this team the team members will have the disciplines product manager, designer, manufacture engineer and product support engineer, with possible Belbin roles as: coordinator, plant, shaper, team worker and completer-finisher and or implementer.

3. For integrated product development the three main phases are: Product Creation (defining and designing a new product), Process Creation (making a new product) and Implementing and Operating (how to use a new product).
   In all teams over the whole Life Cycle these disciplines must be represented. But during the process of development the accents will change and so does the role of the coordinator. During product creation the coordinator will come from a product creation department, and during the process creation from a manufacturing department and during the installation and commissioning from the product support department. Per phase the sort of specialist can also change.

4. For the building environment the mean set up of the teams can be the same as for the product development process, but in this case the team members will come from the different companies as a member of a building consortium (consisting of the participating companies). Also in the beginning the coordinator come from one of the creation firms, during the building phase from one of the executing firms and during the commissioning from one of the building services firms.

Extra Exercises
1. Apply the Belbin methodology for your own team.
2. Investigate your own Belbin role. What are for you the most interesting facts?
3. How will you apply or introduce the concurrent engineering principles in your own environment?

Chapter 3

1. a. Hamburger model of a coffee machine (see figure 8.2).

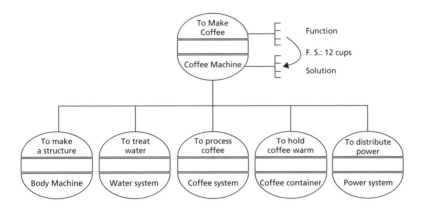

Figure 8.2 Hamburger model of a coffee machine

b. Coffee machine with IDEF-0 (see figure 8.3)
In a coffee machine a lot of processes take place: water intake and water heating, placing filter bag in holder, intake coffee powder (with the weighed amount of coffee powder) in filter bag, bringing hot water to the filter element, collect the fresh coffee in a container, hold coffee warm in this container.

On the second level (A1, A2, A3, etc.) the specifications are as follows:
*   A1 the water process to make hot water (1 to 12 cups), weak to strong
*   A2 the coffee process (1 to 12 cups)
*   A3 the warm hold process of the processed hot coffee (can hold 12 cups of coffee on temperature for one hour, in a container)

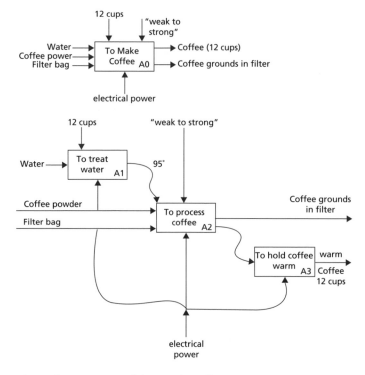

Figure 8.3 IDEF-0 model 'to make coffee'

Chapter 4

1. A delighter for the function 'make coffee' might be:

- level control on the isolated coffee container;
- measure instrument to indicate 'how strong is my coffee'.

2. The Hamburger model of a Kart is illustrated in figure 8.4.

Functional specifications can be:

- Cost price: below €....;
- Maximum speed: 90 km/hour;
- Acceleration : up to 90 km in 5 seconds;
- Maximum fuel consumption: x liter/100 km;
- Safety:

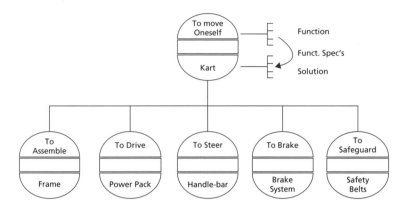

Figure 8.4 Hamburger model for a Kart

3. For making a zero emission (CO_2) power set two options are investigated:

- Electrical power by a fuel cell.
- Mechanical power by a pressed air motor.

The following client wishes are noticed:
- easy to refuel the system;
- easy to drive;
- can drive over minimum of 100 km;
- save to handle.

For these four wishes we develop the planning matrix as part of the HoQ (Voice of the Customer). The next factors have been chosen to fill this planning matrix:

Client wish	WIC	CSP	Goal	ImpR	S.P	RW	NRW	CNRW
Easy to refuel	66	4.2	4.2	1.0	1.0	277.2	0.178	0.178
Easy to drive	85	4.3	4.7	1.12	1.3	532.2	0.343	0.521
Can drive 100 km	75	3.5	4.4	1.25	1.5	492.2	0.317	0.838
Save to handle	50	4.2	4.2	1.0	1.2	252.0	0.162	1.000
Totals						1553.6	1.000	

Table 8.1 HoQ Planning Matrix

The client wishes with the highest score:

- easy to drive (34.3 %);
- can drive over 100 km (31.7%).

Some remarks on the two solutions fuel cell and compressed air motor:

Fuel Cell

Pro's:

- very easy to drive, with high torque at zero speed;
- extra acceleration power by reuse of brake energy (by electrical condenser).

Con's:

- heavy assembly;
- new high technology.

Compressed air motor

Pro's:

- easy technology by combining existing techniques;
- drive like existing fuel engines.

Con's:
- concept not really proven;
- cannot drive over 100 km.

Extra Exercises

1. Make a planning matrix (HoQ) of client wishes for the next building processes in a house:
- to sleep;
- to cook;
- to recreate.

2. What might be a delighter for:

- a bike;
- a CO_2 emission free house.

3. What means creating a CO_2 living environment (HoQ, Planning Matrix, alternative installation concepts)?

Chapter 5
1. The hydraulic system is illustrated in figure 8.5, the Hamburger Model in figure 8.6, the *Fc*-FMEA in table 8.2, and the FTA in figure 8.7.

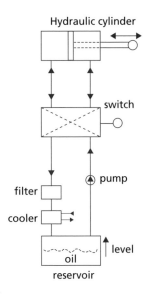

Figure 8.5 Scheme hydraulic cylinder unit

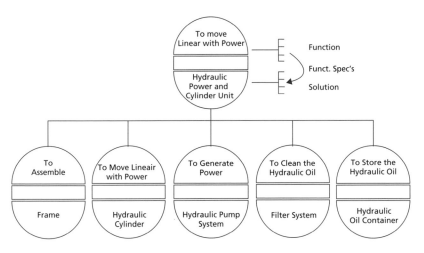

Figure 8.6 Hamburger model hydraulic cylinder unit

FU	FV	FS	FF	KFF	Part	FM	FE
1	A	1	1	1	1	1	1
To move linear power	Hydr power unit	>= F N	< FN	Less Crit	Cylinder	Seal leakage	Less force
					2	2	
					Pipeline	Pipe leakage	Less force
					3	3	
					Pump	Internal leakage	Less force
		2	2	2		1,2,3	
		>= 1 m/min.	< 1 m/min.	Less Crit.	1,2,3	Leakages 1,2,3	Less movement
			3	3	3	3	2
			No move	Critical	Pump	No electrical power	No movement
					4	4	
					Switch	Switch blocked	No movement
					5	5	
					Filter	Filter blocked	No movement.
					6	6	3
					Oil	No oil in container	No pump working so no movement

Table 8.2 Fc-FMEA

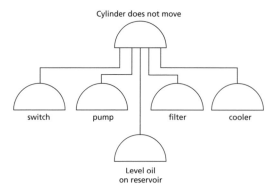

Figure 8.7 FTA hydraulic cylinder unit

2. Switch A:

12000/3000 = 4 times changing a switch = $4 \times 5 + 4 \times 5 = €40$

Extra energy cost over 12000 hrs is = $12000 \times 0.2 \times 0.01 = €24$

Total: €64

Switch B

12000/6000 = 2 times changing a switch = $2 \times 16 + 2 \times 5 = €42$

3.

Total Score:		A		D
• weight:	$1 \times (100)$	$1 \times 50 = 50$		$1 \times 50 = 50$
• well known relation:	$3 \times (100)$	$3 \times 70 = 210$		$3 \times 30 = 90$
• robust to operate:	$2 \times (100)$	$2 \times 50 = 100$		$2 \times 50 = 100$
• just in time delivery	$4 \times (100)$	$4 \times 40 = 160$		$4 \times 60 = 240$
Total:			= 520	= 480

Conclusion: Part A is the best.

4. a. The product configuration model for a pump system family is illustrated in figure 8.8.

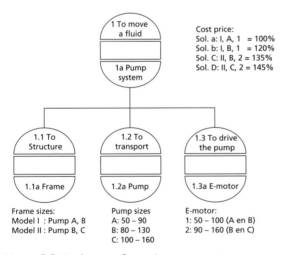

Figure 8.8 Product configuration pump system

Function /Solution	1.1a	1.2a	1.3a	Cost price
1:a cap. A (50-90)	width I	Pump A	Motor 1	100%
cap. B (80-130)	width I	Pump B	Motor 1	120%
	width II	Pump B	Motor 2	135%
cap. C (100-160)	width II	Pump C	Motor 2	145%

b. LCC Calculation for pumps A and B meeting the same functional specifications.

	Type A	Type B
Investment costs	€55000	€50000
Energy costs (4 resp. 4.5% of investment /yr) × 15 yrs	€33000	€33750
Maintenance costs (3.5 resp. 4% of investment/yr) ×15 yrs	€28875	€30000
End of life (10%)	€5500	€5000
Total:	€111375	€108750

Extra Exercises
1. Make a LCC calculation for an old fashion heating and cooling system versus a modern heat pump system.
2. Make a FMEA for an airco system or a heat pump system.

Chapter 6

Step 1: Describe the problem and the problem specifications.
The problem is that in one of the six tanks the temperature of the chocolate does not come higher than 40 °C.
Step 2: Ask ´why´ to formulate a possible cause.
In a chocolate factory 6 new chocolate storage tanks are installed in the newly built hall. The temperature of the chocolate in the tanks has to be at least 50 °C and can vary between 50 and 60 °C. The tanks are heated by a hot water system controlled by a 3-way valve with an entry temperature of 75 °C and an outlet temperature of about 65 °C.
After the first run of starting up the heating system of these tanks, it appeared that in one of the tanks the temperature of the chocolate rose very slowly and that the required temperature of 50 °C was not reached.
What is happening here, and why?

A possible cause is that the temperature control unit is not working properly.

Step 3: Register the possible cause(s) in the table.
Possible causes are:

- the temperature measurement (instrument);
- the 3-way valve;
- the controller.

The outcomes are summarized in table 6.3.

W1	T1	W2	T2	W3	T3	W4	T4	W5	T5
Temp. meas	X								
3-way valve	T	Water flow	T	Water blocked	T	Replace valve	T		
Controller	X								

Table 6.3 Register the Five Why questions

The cause of the problem was the 3-way valve (to control the flow) that blocked the water entry (no hot water flow through the valve) due to the remains of a plastic protecting cap.

Step 4: Check the possible cause(s) to the problem specifications or do a test.
When the outcome of the check or test is True, give the cause a **T** in the register and go to step 5.
When the outcome is not True give the cause a **X** and go back to step 2 (another possible cause).
Step 5: Repeat the steps 1 to 4 five times.

2. In a business proposal there are in general three elements crucial for the go / no go decision:

- market size;
- development costs and time;
- ROI or profitability.

First we have to investigate how big the market is or which percentage of our clients are interested. If a reasonable amount of clients is interested than we investigate the development costs and development time.
We calculate by a break even analysis and a ROI if we are going to decide to invest in the development of this proposal for an electronic speed controller.
The next figures are use for the calculation:

Market Size
25% of all clients have real interest, total amount at least 1000 units, selling 250 units a year.

Development Costs
€120000, so over 4 years €30000/year.
Cost price: €250/unit, Selling price: €500/unit, Net Profit: €60/unit.

Profitability
Client profitability is ROI of 300/500 x 100% = 60% (or within two years).
Break even analysis: 20000/250 = 480 units (far less than 1000).
ROI: (60 x 250 / 30000) x 100% = 50% (over at least 4 years).

Conclusion
For both the pump clients as for the manufacturer the development of this extra device is very profitable.

3. The climate installation did not work properly in that way that the temperature distribution over the metro train was not stable, so that at warm outside temperatures in some section of the metro train the temperature was only 1 or 2 degrees below the outside temperature instead of 6 to 7 degrees.

A brain storm session was organized and at the end of the day two main causes of the problem were pinpointed and chosen for further investigation:

*   cooling capacity of the compressor;
*   flow distribution of the metro train was not adequate.

Control of the compressor capacity (by test) showed no real deviation with the design values.
Control of the flow distribution showed a great deficit of flow, the real value was far below the design values and each train had different positions in the air distribution valves. Further investigation showed (at special tests) that the flow from the air ventilator was far too low. The reason of this low flow was a miss match of driving pulleys for the ventilator so that it's revolutions per minute was too low. By mounting the right set of pulleys and by resetting of the positions of the distribution valves in the train the problem was solved.

Extra Exercises

1. Apply the Fish Bone diagram and the RCRA for the chocolate tempering tank example.
2. Investigate a case in your own environment with a problem of a deviation of a norm and try to apply one of more techniques for solving this problem.
3. Make a business case for an opportunity in your own surroundings.

Chapter 7

1. $A_R = (1700) / (1700 + 52) = 1700 / 1752 = 0.970 = 97.0\%$

$\lambda_R = 8760 / 1752 = 5$ times/year

$TDT_R = 5 \times 52 = 260$ hrs/year

$TPT_R = TOT - TDT_R = 8760 - 260 = 8500$ hours/year

If $MTBF = 167$ hrs and $MTTR_F = 3$ hr, then:

$A_F = (167) / (167 + 3) = 166 / 170 = 0.976 = 97.6\%$

$\lambda_F = (1700 / 170) = 10$ times in 1700 hrs

$TDT_F = 5$ times $\times$ (10 times $\times$ 3) = 150 hours/year

$TPT_F = 8500 - 150 = 8350$ hours/year

$A_{TOT} = A_R \times A_F = 0.970 \times 0.976 = 0.947 = 94.7\%$
$\lambda_{TOT} = \lambda_R + \lambda_R \times \lambda_F = 5 + 5 \times 10 = 55$ times stop/year
$TDT_{TOT} = 260 + 150 = 410$ hours/year
$TPT_{TOT} = TOT - TDT_{TOT} = 8760 - 410 = 8350$ hours/year

2. See also figure 8.9.

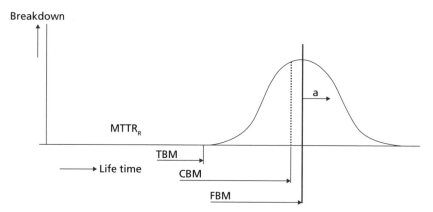

Figure 8.9

- Cylinder liner of a piston compressor:
  - regular wear pattern: TPM.
- Roll bearing of a centrifugal compressor:
  - wear pattern depending on operating context: CBM.
- Filter set of filtering waste water:
  - Not very important, use pattern known: FBM.
- A lot of lamps in a big office:
  - Wear pattern with narrow distribution: TPM , by changing all lamps at once, or FBM by changing each lamp apart.
- Tyres of busses in a transport company:
  - Regular wear pattern, style of driving more or less narrow: TPM.

3. CBM: $5 \times (2 \times 175 + 3 \times 50 + 1 \times 100) = €3000 / 3$ years
TBM: $7 \times (2 \times 175 + 1 \times 100) = €3150 / 3$ years
CBM is the cheapest strategy.

4. See table 8.4

FU	SO	FS	FF	CM	Ranking	FM	FE
1	A	1	A	1		1	
Trans-port Air	Centr.. Comp	Q =1000 m³/hr and Δh = 500 mm H$_2$O	Q< 1000 (and or Δh< 500)	FS3.4	Task Sel.	Impellor wears	Less throughput
						2	
						Seals leakage	Less throughput .
			B	2		1	
			Q=0, Δh=0	FS4.3	Task Sel.	No power	No throughput
						2	
						Bearing damage	No throughput.
						Suction side	
						3	
						Bearing damage	No throughput
						Pression side	

Table 8.4   Fc-FMEA

Task selection

Asset or Part	Ques. 1	Ques. 2	Ques. 3	Ques. 4	Ques. 5	Ques. 6	Repair strategy
Impellor	Yes	Yes	Yes	Yes	Yes	-	CBM

5. Availability $A = (120-15)/120 = 0.875$
Productivity P = net capacity / standard capacity = (net capacity/set up capacity) × (set up capacity/standard capacity)
Net capacity = $21.0 / (120 - 15) = 0.20$ ton/hrs
So Productivity $P = (0.20 / 0.22) \times (0.22 / 0.25) = 0.800$
Quality $Q = (21.0 - 2.0) /21.0 = 0.905$
OEE = $A \times P \times Q = 0.875 \times 0.800 \times 0.905 = 0.6335 = 63.4\%$

Extra Exercises

1. Develop an integrated maintenance plan (*Fc*-RCM, *Fc*-FMEA, FTA, Task selection) for a product or process in your own working environment.
2. Make a *Fc*-FMEA of an airco system (heating and ventilation) for a small office.

Index

3 RP Fields for a successfully
    Business Process 8
80/20 rule 173

A

after sales business activity 137
after sales engineering. 139, 152
After Sales Problems 136
Assembly to Order. 121
autonomous maintenance 191, 195
Availability 193

B

basis routines, problem solving 143
Belbin team roles. 22
benchmarks 81
benchmarks, functionality 81
benchmarks, performance 81
brainstorming .142
Business Model for Improvements . . . 152

C

cause-effect diagram 149
chambers of the HoQ 74
Clausing, D. 110
Client's Quality Characteristics. 74
client thinking 62
client wishes. 61
coaching 29
code systems 50

code systems for product
    development teams 55
collaborative engineering 49
commissioning .189
competitive satisfaction
    performance. 69
completer-finisher 22
concept selection, Pugh
    methodology. 111
concept selection, van den
    Kroonenberg methodology 112
concurrent engineering. 49
Concurrent Engineering 30
Concurrent Engineering Planning. . .85, 97
Condition-Based Maintenance 168
Contradiction 108
coordinator. 22
corrective maintenance 173
cost elements, manufacturing 123
cost engineering. 126
Costs of Manufacturing. 99
Cradle to Cradle 123
creating business with clients. 61
creator. 22
criticality. 104
criticality matrix. 104
Criticality Matrix 181
crowd sourcing. 96
cumulative normalized raw weight . . . 70
customer satisfaction. 72

D

degradation process 161, 167
delighters 73
Design for Assembling 119
design for continuous improvement . .139
Design for Life Cycle Costs 123
Design for Long Life Service 122
Design for Production Costs 123
Design for Manufacturing 120
Design for Operation and
  Maintainabilty 121
Design for Reuse 123
Design for Sustainability 123
designing improvements 141
Design Kernel for Engineering 126
design phase 91
design requirements 74
discipline matrix 28

E

early equipment flows 190
Early Equipment Management 189
eliminating complaints 140
eliminating complaints,
  methodologies 141
end of life phase 96
Engineering to Order 121
Enhanced Quality Function Deployment
  (EQFD) 110
Enterprise Resource Planning (ERP) . . 98
ergonomics 121
evident function 182
Evolution 108

F

fact person 26
failure 179
Failure-Based Maintenance 171
failure consequences 182
Failure Consequences Analysis
  (FCA) 182
failure consequences, ranking 183

failure effects .180
Failure Elimination Modification 173
failure mode 180
Failure Modes and Effects Analysis . . .102
failure tree 150
Fault Tree Analysis (FTA) 106
Fc-FMEA 104
Fc-RCM 186
fishbone diagram 149
five levels of innovation 109
Five Times Why Question analysis 149
flexible person .27
FMEA model 102
full Fc-RCM process 186
Functional critical Failure Modes and
  Effects Analysis (Fc-FMEA) 180
Functional critical Reliability Centred
  Maintenance (Fc-RCM) 176
functional decomposition 179
functional decomposition model 46
functional failure 179
Functionality .108
functional specifications 177
functions 61

H

Hamburger model 46, 136, 177
Hamburger model,
  Function-Solutions 116
Hamburger model, planning 99
hidden factory 194
hidden failure .182
hidden function 182
hierarchical breakdown model 51
historical data .43
House of Quality 68
House of Quality, Pugh 111

I

ICT implementation failures 41
Ideality 108
IDE as a Business Process 8

IDE characteristics 9
IDE company model 12, 60
IDE competences. 14, 24
IDE director roles. 25
IDE domain 10
IDE and ICT. 11
IDE kernel 10
IDE product development team 28
IDE space 10
IDE structure models 43
IDE team 21
IDEF-0 model 44
implementation ICT processes. 42
implementer. .22
Improvement Activities. 140
Improvement-Based Maintenance 172
improvement ratio. 69
installation phase. 95
Integrated Design and Engineering,
    business process. 124
Integration Definition Function Modeling
    (IDEF) 44
introvert person 26
investment costst calculation 101
Ishikawa diagram 149

K

Kano model 72
Kano model and key SQC's. 82
kernel of IDE .10
Kepner and Tregoe methodology 143
key SQC's. 81
Kroonenberg selection methodology . . 112

L

Lean Process and Product
    Development 31
Lean Process Methodology 31
Lean product development 125
Life Cycle 90
Life Cycle and Investment. 99
Life Cycle Costs 124

Life Cycle Costs (LCC). 100

M

maintainability. 121
Maintenance. 160
maintenance degradation process . . . 166
maintenance need 160
maintenance plan, product
    development. 189
Maintenance strategies, new
    products 175
make or buy decision. 119
Management Paradox 35, 56
manufacturing costs 99
Manufacturing Selection. 119
mapping client wishes. 66
mass consumer manufacturing, after
    sales. 138
mass production 119
mean life time .167
Mean Time Between Failures
    (MTBF) 114
mean time between repairs. 162
mean time to repair. 162
methodology of Kepner and Tregoe . . . 143
modification. 172
modification by eliminating
    failures 173
monitor-evaluator. 23
Morphological Scheme 112
multidisciplinary development team . . .23
multi disciplinary teams, TPM 191
Myers Briggs Type Indicator. 26

N

need for maintenance 176
noises 114
normal distribution 167
normalized raw weight 70

O

objective person. 27

operating context 168
operation. 121
operational phase. 96
out of function phase. 96
outsourcing 119
Overall Equipment Efficiency
    (OEE). 193

P

Pareto analysis 142
Pareto diagram. 173
Part Engineering 107
Parts Engineering 102
Pay Back Period 101
Planned Preventive Maintenance
    (PPM) 161
Planning Matrix. 69
plant 22
portfolio matrix 6
possibility person. 26
preventive maintenance 170
primary functions 177
priorities. 76
priorities, key SQC's 81
Problem Specification table 146
process driven manufacturing 119
Product Configuration Model. 55, 128
Product Definition Document
    (PDFDoc) 84
Product Design Document
    (PDSDoc) 92
Product Engineering Document
    (PENDoc) 94
Product Life Cycle 5
Production Efficiency 193
Production Losses 193
production/manufacturing cost 123
Product Manufacturing Document
    (PMADoc). 94

Q

Quality Efficiency. 193
Quality Function Deployment (QFD). . 68
Quality Losses .193

R

RAMSHE specifications 92
RAM specifications 75
raw weight 70
raw weight, cumulative normalized . . . 70
raw weight, normalized. 70
RCFA methodology 150
relation customer needs and SQC's. . . 76
relation matrix .55
reliability behavior 163
Reliability Centred Maintenance
    (RCM) 176
reliability (R). .163
Repair after Failure Breakdown
    (RFB). 173
repair strategie 166
requirements 114
Resource. 108
resource-investigator 22
responsibility matrix 80
responsibility symbols. 80
Return on Investment (ROI) 2
reuse 123
R.M. Belbin 22
robustness 114
Root Cause Failure Analysis (RCFA) . .150
run to failure 171

S

sales point. 69
satisfiers 73
secondary functions. 177
selection matrix 114
selection matrix, Pugh. 111
self perception matrix 23

shaper 22
simultaneous engineering 49
single product production 119
six big losses 195
specialist 23
Structure Models for IDE
    Applications 43
structurist 55
subjective person 27
substitute quality characteristics 74
Substitute Quality Characteristics
    (SQC) 69
sustainability 123
Systems engineering 49

T
Taguchi, G. 114
Taguchi methodology 114
targets for key SQC's 82
Task Selection Scheme 184
team requirements 29
team roles 22
team worker 22
technical relationships 78
technical response 74

Time-Based Maintenance 169
Total Dead Time (TDT) 162
Total Operational Time (TOT) 162
Total Productive Maintenance (TPM) . . 191
TRIZ 108
TRIZ key elements 108

U
Unplanned Failure Maintenance
    (UFM) 165
usefull operational time 193

V
value 124
value chain 125
value creation process 124
V-model 49
Voice of the Customer 69
Voice of the Developer 74

W
wear 167
WEBIM model 98
wishes 114